Modern Computational Methods for Fractional Differential Equations

This book provides a comprehensive introduction to the fundamentals of fractional calculus. Covering a range of analytical and numerical methods specifically designed for fractional calculus problems, this book explains the step-by-step implementation and computational aspects of these methods. Readers of this book should feel empowered to effectively apply analytical and numerical techniques to solve fractional calculus problems.

Features

- Practical examples and case studies
- Suitable material for professionals and postgraduate researchers
- Numerous interesting and novel problems

Modern Computational Methods for Fractional Differential Equations

Hossein Jafari

University of Mazandaran
Babolsar, Iran
University of South Africa
UNISA0003, South Africa

Haleh Tajadodi

University of Sistan and Baluchestan
Zahedan, Iran

Yusif S. Gasimov

Azerbaijan University
Baku, Azerbaijan

CRC Press
Taylor & Francis Group
Boca Raton London New York

CRC Press is an imprint of the
Taylor & Francis Group, an **informa** business
A CHAPMAN & HALL BOOK

Designed cover image: Shutterstock 2384032727

First edition published 2025
by CRC Press
2385 NW Executive Center Drive, Suite 320, Boca Raton FL 33431

and by CRC Press
4 Park Square, Milton Park, Abingdon, Oxon, OX14 4RN

CRC Press is an imprint of Taylor & Francis Group, LLC

Library of Congress Cataloging-in-Publication Data
Names: Jafari, Hossein (Professor) author. | Tajadodi, Haleh, 1984-, author. | Gasimov, Yusif, author.
Title: Modern computational methods for fractional differential equations / Hossein Jafari, University of South Africa. UNISA0003, South Africa, University of Mazandaran, Babolsar, Iran, Haleh Tajadodi, University of Sistan and Baluchestan, Zahedan, Iran, Yusif Gasimov, Azerbaijan University, Baku, Azerbaijan.
Description: First edition. | Boca Raton : C&H/CRC Press, 2025. |
Includes bibliographical references and index.
Identifiers: LCCN 2024046886 (print) | LCCN 2024046887 (ebook) |
ISBN 9781032750439 (hardback) | ISBN 9781032755731 (paperback) |
ISBN 9781003474609 (ebook)
Subjects: LCSH: Integral equations. | Calculus, Operational. | Mathematical models.
Classification: LCC QA431 .J335 2025 (print) | LCC QA431 (ebook) |
DDC 515/.35–dc23/eng/20250102
LC record available at https://lccn.loc.gov/2024046886
LC ebook record available at https://lccn.loc.gov/2024046887

ISBN: 9781032750439 (hbk)
ISBN: 9781032755731 (pbk)
ISBN: 9781003474609 (ebk)

DOI: 10.1201/9781003474609

Typeset in LMRoman
by codeMantra

We dedicate this book to
our family members

Contents

Preface

FRACTIONAL CALCULUS with a long history has many applications in the real world. There are several types of fractional derivatives such as Grünwald-Letnikov, Riemann-Liouville, Caputo, Caputo-Fabrizio and Atangana-Baleanu. However, all the problems in this book deal with fractional differential equations and some fractional derivatives such as the Riemann-Liouville and Caputo fractional derivatives. In the last few decades, various semi-analytical and numerical methods have been used for solving fractional-order differential equations.

This book contains seven chapters. Several analytical/approximate methods such as the Adomian decomposition method, homotopy perturbation method, homotopy analysis method, differential transform method and variational iteration method are presented to obtain exact/approximate solution of fractional differential equations, fractional integral equations and fractional partial differential equations. Also, several numerical methods such as Adams-Bashforth-Moulton method, operational matrices and wavelet method are used to find numerical solutions of various types of fractional-order differential equations.

This book will be useful to scientists and postgraduate students doing research in the field of fractional calculus.

Hossein Jafari
Haleh Tajadodi
Yusif S. Gasimov

Author Bios

Prof. Hossein Jafari is a Full Professor of Applied Mathematics. He completed his Ph.D. in Mathematics in Savitribai Phule Pune University, formerly the University of Pune, India, under the supervision of Prof. Varsha Daftardar-Gejji. His research interests include numerical analysis, Lie symmetry, integral transforms, bio-mathematics, artificial neural network and approximation theory. He gave a general integral transform for solving ODEs and integral equations. Since 2017, he has been on the list of the Essential Science Indicators (ESI) as a top researcher by Web of Science.

Dr. Haleh Tajadodi earned her Ph.D. from the University of Mazandaran, Iran, in 2014. She is an Associate Professor in the Department of Mathematics, University of Sistan and Baluchestan. Her research focused on fractional calculus. She studied numerical and analytical methods for solutions of fractional differential equations. Additionally, her research interests include spectral method and approximation theory.

Prof. Yusif S. Gasimov graduated from Baku State University in 1992. He earned his Ph.D. in Mathematical Analysis in 1997 and Doctor of Sciences in Numerical Methods in 2011.

Y. S. Gasimov developed methods to study eigenvalue problems with variable domains and shape optimization problems. He proposed a mathematically constructive formula for the definition of the domain variation, which plays a decisive role in the investigation of shape optimization problems. He gave a formulation of the inverse spectral problem with respect to domain, introduced the concept of s-functions and proposed a method for solution of this problem by the given set of these functions.

Yusif S. Gasimov also studied numerical and analytical methods for solutions of different types of direct and inverse equations and also fractional derivative equations.

CHAPTER 1

Preliminaries

THIS CHAPTER begins with an introduction to the history of establishing fractional calculus generally. Then, the basic definitions of fractional calculus as well as fractional integrals and derivatives introduced by different authors that will be employed in the next chapters are given. Here the gamma function and Mittag-Leffler function, and then Grünwald-Letnikov's, Riemann-Liouville and Caputo fractional derivatives are presented.

1.1 HISTORY

Fractional calculus deals with generalizations of the ordinary differentiation and integration to non-integer (real/complex) orders. This subject is as old as the calculus of differentiation and goes back to the times of Leibniz, Gauss and Newton. The first reported attempt to generalize derivatives to fractional order is contained in the correspondence of Leibniz (1695) with L'Hospital [81]. In a letter to L'Hospital in 1695, Leibniz raised the following question: "Can the meaning of derivatives with integer order be generalized to derivatives with non-integer orders?" The story goes that L'Hospital was somewhat curious about that question and replied by another question to Leibniz. "What if the order will be 1/2?" Leibniz in a letter dated September 30, 1695 replied: "It will lead to a paradox, from which one day useful consequences will be drawn." In these words, fractional calculus was born. Following L'Hospital's and Leibniz's first inquisition, fractional calculus was primarily a study reserved for the best minds in mathematics. Fourier, Euler and Laplace are among the many who dabbled with fractional calculus and the mathematical consequences [85]. Mathematicians like Abel, Gauss, Fourier, Grünwald, Riemann, Liouville, Weyl, and Letnikov made major contributions to the subject of fractional calculus. Abel was the first person who applied fractional derivative for solving a generalized version of the tautochrone problem. He solved the following integral equation for $\alpha = -\frac{1}{2}$,

$$k = \int_0^x (x-t)^\alpha f(t)dt, \tag{1.1}$$

DOI: 10.1201/9781003474609-1

where $f(t)$ is unknown. For determining f, Abel wrote the right-hand side of Eq. (1.1) as $\sqrt{\pi}\left[\frac{d^{-1/2}f(x)}{dx^{-1/2}}\right]$ and applied $\frac{d^{-1/2}f(x)}{dx^{-1/2}}$ on both sides of Eq. (1.1) to obtain:

$$\frac{d^{1/2}k}{dx^{1/2}} = \sqrt{\pi}f(x),$$

as the fractional operators (under suitable conditions on f) have the property:

$$\frac{d^{1/2}}{dx^{1/2}}\left[\frac{d^{-1/2}f(x)}{dx^{-1/2}}\right] = \frac{d^0 f(x)}{dx^0} = f(x).$$

This is a remarkable achievement of Abel, which gave impetus to the development of fractional calculus. He further applied the theory of fractional calculus and applied it successfully to problems in potential theory [81]. Many definitions that fit the concept of a non-integer order integral or derivative exist in the literature. The most famous of these definitions in the world of fractional calculus are the Riemann-Liouville, Grünwald-Letnikov and more recently Caputo definitions. Caputo reformulated the 'classic' definition of the Riemann-Liouville fractional derivative in order to use integer order initial conditions to solve his fractional order differential equations [92]. This subject has gained importance and popularity during the past three decades or so, due mainly to its demonstrated applications in numerous and seemingly diverse fields of science and engineering [45,92]. It indeed provides several potentially useful tools for solving differential and integral equations. Most of the mathematical theory applicable to the study of fractional calculus was developed prior to the turn of the 20th century. However, it is in the past 100 years that the most intriguing leaps in engineering and scientific application have been found.

There are many fields of applications where we can use the fractional calculus, such as Viscoelasticity, Control theory, Heat conduction, Electricity, Mechanics, Chaos, Fractals, etc. [32, 59, 77]. Recently, Maxwell's equations have been generalized using fractional derivatives to better understand multipole moments [27, 28]. Riewe [95,96] has formulated Lagrangian and Hamiltonian mechanics involving fractional derivatives which leads to equations of motion with non-conservative forces such as friction. Further, Agrawal [3] has developed calculus of variations including fractional order derivatives. Many structures found in nature can be modeled by fractals [34, 78, 113]. Fractals are often so irregular that methods of ordinary calculus are either inapplicable or ineffective. Fractional calculus has proved effective in formulating processes involving fractal structures and phenomenon [9, 71, 89]. Fractional calculus is also being applied to statistics [90, 91].

1.2 BASIC DEFINITIONS

1.2.1 Gamma Function

Historically, the gamma function was introduced by L. Euler (1707–1783) in 1729 [5].

Definition 1.1 *The gamma function is defined as*

$$\Gamma(z) = \int_0^\infty e^{-t} t^{z-1}\, dt, \qquad \mathrm{Re}(z) > 0. \tag{1.2}$$

It can be shown that this integral converges uniformly for all values of z for which $\mathrm{Re}(z) > 0$.

The function $\Gamma(z)$ is bounded and differentiable and in fact, is analytic throughout the domain. By performing integration by parts, we obtain

$$\Gamma(z+1) = \int_0^\infty e^{-t} t^{z}\, dt = e^{-t} t^{z}|_0^\infty + z \int_0^\infty e^{-t} t^{z-1}\, dt,$$

and deduce the simple but important recurrence formula

$$\Gamma(z+1) = z\,\Gamma(z). \tag{1.3}$$

Analytic Continuation for Re$(z) < 0$

An analytic continuation of $\Gamma(z)$ to the left of imaginary axis can be accomplished through the repeated use of the recurrent formula (1.3) expressed in the form

$$\Gamma(z) = \frac{\Gamma(z+1)}{z}, \quad z \neq 0. \tag{1.4}$$

The right-hand side of Eq. (1.4) is defined for all $\mathrm{Re}(z) > -1$, $z \neq 0$, and thus this expression defines $\Gamma(z)$ in this domain. Replacing z by $z+1$ in Eq. (1.4) yields

$$\Gamma(z+1) = \frac{\Gamma(z+2)}{(z+1)}, \quad z \neq -1,$$

and when substituted into Eq. (1.4), we obtain

$$\Gamma(z) = \frac{\Gamma(z+2)}{z(z+1)}, \quad z \neq 0,\ -1,$$

which now defines $\Gamma(z)$ for all $\mathrm{Re}(z) > -2$, $z \neq 0,\ -1$. Continuing this process, we deduce that

$$\Gamma(z) = \frac{\Gamma(z+k)}{z(z+1)\cdots(z+k-1)}, \quad z \neq 0,\ -1, \ldots, -k+1, \tag{1.5}$$

where k is a positive integer. Equation (1.5) can be used to define the gamma function at every z with a negative real part except at negative integers and zero. The values $z = 0,\ -1,\ -2, \ldots,$ are actually first-order*poles* of the function and thus

$$|\Gamma(-n)| = \infty, \quad n = 0,\ 1, \ldots.$$

The graph of the gamma function for $z = x$, a real variable, is sketched in Figure 1.1.

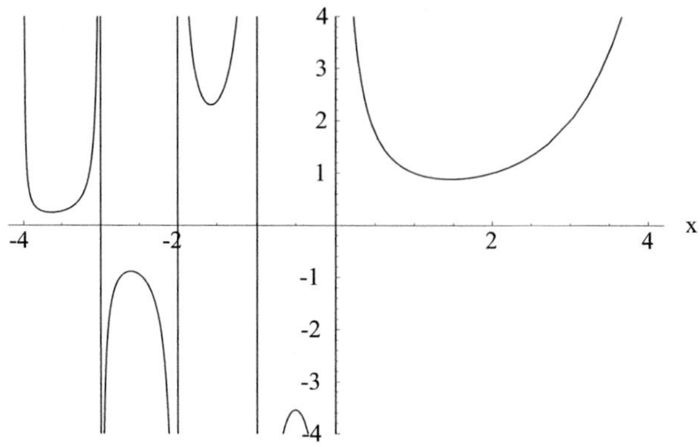

Figure 1.1 Gamma function.

Basic identities for $\Gamma(z)$

$$\begin{aligned}
&\Gamma(z) = \int_0^{\infty} e^{-t} t^{z-1}\, dt, \qquad Re(z) > 0,\\
&\Gamma(z+1) = z\,\Gamma(z),\\
&\Gamma(z) = \frac{\Gamma(z+k)}{z(z+1)\cdots(z+k-1)}, \quad k = 1,\, 2, \ldots, \\
&\Gamma(z)\Gamma(1-z) = \frac{\pi}{\sin \pi z}, \qquad z \text{ nonintegral},\\
&\Gamma(n+1) = n!, \qquad n = 0,\, 1,\, 2, \ldots,\\
&\Gamma(n+1) \approx \sqrt{2\,\pi\, n} n^n\, e^{-n}, \quad n \longrightarrow \infty, \text{(Stirling's formula)}.
\end{aligned} \tag{1.6}$$

1.2.2 Mittag-Leffler Function

The Mittag-Leffler function is an important function that finds widespread use in the field of fractional calculus. It is a generalization of the exponential function. Just as the exponential function naturally arises in the solution of integer order differential equations, the Mittag-Leffler function plays an analogous role in the solution of non-integer order differential equations.

Definition 1.2 *The standard definition of the Mittag-Leffler function [92] is*

$$E_\alpha(z) = \sum_{k=0}^{\infty} \frac{z^k}{\Gamma(\alpha\, k + 1)}, \quad \alpha > 0. \tag{1.7}$$

The exponential function corresponds to $\alpha = 1$ in the expression (1.7).

Figure 1.2 illustrates the Mittag-Leffler function for $\alpha = 1,\, 1.5,\, 2,\, 5$.

A two-parameter Mittag-Leffler function is defined by the series expansion [92]

$$E_{\alpha,\beta}(z) = \sum_{k=0}^{\infty} \frac{z^k}{\Gamma(\alpha\, k + \beta)}, \quad (\alpha > 0,\, \beta > 0). \tag{1.8}$$

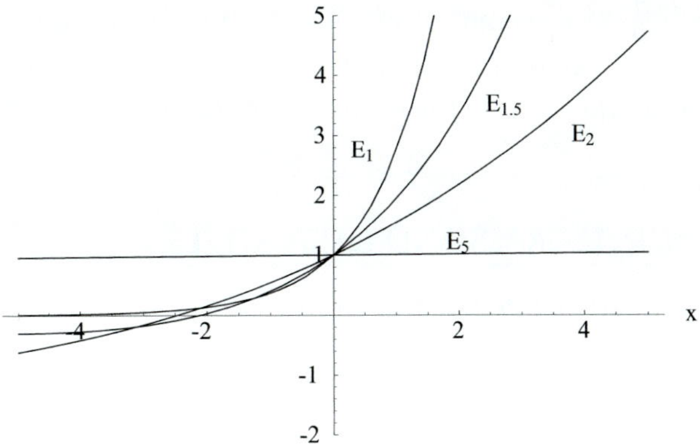

Figure 1.2 Mittag-Leffler function.

It follows from the definition (1.8) that

$$\begin{aligned}
E_{\alpha,1}(z) &= \sum_{k=0}^{\infty} \frac{z^k}{\Gamma(\alpha k+1)} \equiv E_\alpha(z), \\
E_{1,1}(z) &= \sum_{k=0}^{\infty} \frac{z^k}{\Gamma(k+1)} = \sum_{k=0}^{\infty} \frac{z^k}{k!} = e^z, \\
E_{1,2}(z) &= \sum_{k=0}^{\infty} \frac{z^k}{\Gamma(k+2)} = \frac{e^z-1}{z}, \\
E_{1,m}(z) &= \sum_{k=0}^{\infty} \frac{z^k}{\Gamma(k+m)} = \frac{1}{z^{m-1}}\left\{e^z - \sum_{k=0}^{m-2} \frac{z^k}{k!}\right\}.
\end{aligned} \tag{1.9}$$

Multivariate Mittag-Leffler function [92] is defined below:

$$E_{(\alpha_1,\ldots,\alpha_n),\beta}(z_1,z_2,\ldots,z_{n+1}) = \sum_{k=0}^{\infty} \sum_{\substack{l_1+\cdots+l_{n+1}=k \\ l_j\geq 0}} (k;l_1,\ldots,l_{n+1}) \left[\frac{\prod_{j=1}^{n+1} z_j^l}{\Gamma(\beta+\sum_{j=1}^{n+1}\beta_j l_j)}\right].$$

1.2.2.1 Leibnitz Rule for Differentiating an Integral

For differentiation of the integral $\int_{f(x)}^{h(x)} G(x,t)\,dt$ with respect to x, we apply the useful Leibnitz rule given by Ref. [112]:

$$\frac{d}{dx}\int_{f(x)}^{h(x)} G(x,t)\,dt = G(x,h(x))\frac{dh(x)}{dx} - G(x,f(x))\frac{df(x)}{dx} + \int_{f(x)}^{h(x)} \frac{\partial G(x,t)}{\partial x}\,dt, \tag{1.10}$$

where $G(x,t)$ and $\frac{\partial G(x,t)}{\partial x}$ are continuous functions in the domain D in the xt-plane that contains the rectangular region $\mathbb{R}, a \leq x \leq b$, $t_0 \leq t \leq t_1$ and the limits of integration $f(x)$ and $h(x)$ are defined functions having continuous derivative for $a < x < b$.

1.3 FRACTIONAL INTEGRALS AND DERIVATIVES

1.3.1 Grünwald-Letnikov's Integral and Derivative

Let $f \in C[a,b]$. The first-order derivative of the function $f(x)$ is defined as

$$f'(x) = \frac{df}{dx} = \lim_{h \to 0} \frac{f(x+h) - f(x)}{h}, \text{ if it exists.} \tag{1.11}$$

Applying this definition twice we can find the second-order derivative

$$f^{''}(x) = \lim_{h \to 0} \frac{f(x+2h) - 2f(x+h) + f(x)}{h^2}.$$

By induction,

$$f^{(n)}(x) = D^n f(x) = \lim_{h \to 0} \frac{1}{h^n} \sum_{m=0}^{n} (-1)^r \binom{n}{m} f(x - mh), \tag{1.12}$$

where $\binom{n}{m} = \frac{\Gamma(n+1)}{m!\,\Gamma(n-m+1)}$. Expression (1.12) can be generalized for non-integer values of n. For positive values of n, it represents fractional derivative, and for negative values, it represents fractional integral. Let us now consider the following expression where α is non-integer.

$${}^{G}_{a}D^{\alpha}_{x} f(x) = \lim_{h \to 0} \frac{1}{h^{\alpha}} \sum_{m=0}^{\frac{t-a}{h}} (-1)^m \binom{\alpha}{m} f(x - mh), \quad \alpha > 0, \tag{1.13}$$

where $\binom{\alpha}{m} = \frac{\Gamma(\alpha+1)}{m!\,\Gamma(\alpha-m+1)}$ and t and a are the upper and lower limits of differentiation, respectively. Equation (6.124) is called Grünwald-Letnikov (GL) derivative.

Example 1.1

$$\begin{aligned}
{}^{G}_{a}D^{\alpha}_{x} e^{ax} &= \lim_{h \to 0} h^{-\alpha} \sum_{m=0}^{\alpha} (-1)^n \binom{\alpha}{m} e^{a(x+(\alpha-n)h)} \\
&= e^{ax} \lim_{h \to 0} h^{-\alpha} \sum_{m=0}^{\alpha} (-1)^n \binom{\alpha}{m} (e^{ah})^{\alpha-n} \\
&= e^{ax} \lim_{h \to 0} h^{-\alpha} (e^{ah} - 1)^{\alpha} = a^{\alpha}\, e^{ax}.
\end{aligned}$$

The GL derivative can be extended to include negative values of α, using generalized binomial coefficients. Generalized binomial coefficient is defined as:

$$\binom{-\alpha}{m} = (-1)^m \frac{\Gamma(\alpha+m)}{\Gamma(\alpha)\, m!}, \qquad \alpha > 0. \tag{1.14}$$

Using (1.14) we can now extend (6.124) for negative orders, *i.e.*

$${}_a^G D_x^{-\alpha} f(x) = \lim_{h\to\ 0} h^\alpha \sum_{m=0}^{\frac{t-a}{h}} (-1)^m \frac{\Gamma(\alpha+m)}{m!\,\Gamma(\alpha)} f(x-mh), \tag{1.15}$$

which is called GL fractional integral.

Example 1.2 *Let us consider several particular cases.*
For $\alpha = 1$, we have:

$$\begin{aligned} D^{-1} f(t) &= \lim_{\substack{n\to\infty \\ h\to 0}} h \sum_{m=0}^{n} \frac{\Gamma(m+1)}{m!\Gamma(1)} f(t-mh) = \lim_{\substack{n\to\infty \\ h\to 0}} \sum_{m=0}^{n} h f(t-mh) \\ &= \lim_{\substack{n\to\infty \\ h\to 0}} \int_0^{nh} f(t-z)dz = \lim_{\substack{n\to\infty \\ h\to 0}} \int_{t-nh}^{t} f(z)dz = \int_a^t f(z)dz, \quad h = \frac{t-a}{n}, \end{aligned}$$

For $\alpha = 2$, we get

$$\begin{aligned} f_h^{(-2)}(t) &= h^2 \sum_{m=0}^{n} \frac{\Gamma(m+2)}{m!\Gamma(2)} f(t-mh) = h^2 \sum_{m=0}^{n} (m+1) f(t-mh) \\ &= h \sum_{m=1}^{n+1} (mh) f(y-mh), \end{aligned}$$

and taking $h \to 0$ we obtain

$$\lim_{\substack{n\to\infty \\ h\to 0}} f_h^{(-2)}(t) =_a D_t^{-2} f(t) = \int_0^{t-a} z f(t-z)dz = \int_a^t (t-z) f(z)dz = I_a^2 f(t), \tag{1.16}$$

as $y \longrightarrow t$ as $h \longrightarrow 0$.

Using induction, we get the following general expression:

$${}_a^G D_t^\alpha f(t) = \lim_{\substack{n\to\infty \\ h\to 0}} h^\alpha \sum_{m=0}^{n} \frac{\Gamma(m+\alpha)}{m!\Gamma(\alpha)} f(t-mh) = \frac{1}{(\alpha-1)!} \int_a^t (t-z)^{\alpha-1} f(z)dz. \tag{1.17}$$

1.3.2 Riemann-Liouville Fractional Integral

Riemann-Liouville (R-L) fractional integral operator is a direct generalization of Cauchy's formula for an n-fold integral.

$$\underbrace{\int_a^x \int_a^{x_1} \cdots \int_a^{x_{n-1}}}_{n \text{ times}} f(t)\, dt\, dx_{n-1} \ldots dx_1 = \frac{1}{(n-1)!} \int_a^x \frac{f(t)}{(x-t)^{1-n}} dt. \tag{1.18}$$

Definition 1.3 *If $f(x) \in C[a,b]$ and $\alpha > 0$ then*

$$
\begin{aligned}
{}_{a^+}I_x^{\alpha} f(x) &:= \frac{1}{\Gamma(\alpha)} \int_a^x \frac{f(t)}{(x-t)^{1-\alpha}} dt, \quad x > a, \\
{}_{b^-}I_x^{\alpha} f(x) &:= \frac{1}{\Gamma(\alpha)} \int_x^b \frac{f(t)}{(x-t)^{1-\alpha}} dt, \quad x < b,
\end{aligned}
\tag{1.19}
$$

are called as the left-sided and the right-sided Riemann-Liouville fractional integral of order α, respectively.

1.3.3 Riemann-Liouville Fractional Derivative

The definition of Riemann-Liouville fractional derivative of order α is motivated by Abel's integral equation for any $\alpha \in (0,\, 1)$. Consider the integral equation:

$$
f(x) = \frac{1}{\Gamma(\alpha)} \int_a^x \frac{\phi(t)}{(x-t)^{1-\alpha}} dt, \quad x > 0. \tag{1.20}
$$

Equation (1.20) can be solved by multiplying both sides by $(x-a)^{-\alpha}$ and integrating:

$$
\int_a^x \frac{dt}{(x-a)^{\alpha}} \int_a^t \frac{\phi(s)}{(t-s)^{1-\alpha}} ds = \Gamma(\alpha) \int_a^x \frac{f(t)}{(x-t)^{\alpha}} dt. \tag{1.21}
$$

Using Fubini's theorem and interchanging the order of integration in LHS of Eq. (1.21), we get

$$
\int_a^x \phi(s)\, ds \int_s^x \frac{dt}{(x-t)^{\alpha}(t-s)^{1-\alpha}} = \Gamma(\alpha) \int_a^x \frac{f(t)}{(x-t)^{\alpha}} dt. \tag{1.22}
$$

The inner integral is evaluated by substituting $t = s + \tau(x-s)$ and using the formula for beta function,

$$
\int_s^x (x-t)^{-\alpha}(t-s)^{\alpha-1} dt = \int_0^1 (1-\tau)^{-\alpha}(\tau)^{\alpha-1} d\tau = B(\alpha, 1-\alpha). \tag{1.23}
$$

In view of Eq. (1.23), Eq. (1.22) yields

$$
\int_a^x \phi(s)\, ds = \frac{1}{\Gamma(1-\alpha)} \int_a^x \frac{f(t)}{(x-t)^{\alpha}} dt. \tag{1.24}
$$

Differentiating Eq. (1.24), we get

$$
\phi(x) = \frac{1}{\Gamma(1-\alpha)} \frac{d}{dx} \int_a^x \frac{f(t)}{(x-t)^{\alpha}} dt. \tag{1.25}
$$

Thus solution of Eq. (1.20) is given by Eq. (1.25) for $\alpha \in (0,1)$. As Eq. (1.20) is the integral of order α, it is natural to define the inversion (1.25) as derivative of order α.

Definition 1.4 *If $f(x) \in C[a,b]$ and $a < x < b$ then*

$$ {}_a\mathbb{D}^{\alpha_x} f(x) := \frac{1}{\Gamma(1-\alpha)} \frac{d}{dx} \int_a^x \frac{f(t)}{(x-t)^\alpha} dt = D_a I_x^{1-\alpha} f(x), \quad \alpha \in (0,1), \tag{1.26} $$

is called the left side Riemann-Liouville fractional derivative of order α.

In the next definition, we present the Riemann-Liouville fractional derivative for arbitrary value of $\alpha > 0$.

Definition 1.5 *If $f \in C[a,b]$ and $n-1 < \alpha \leq n$ then the left-sided and right-sided Riemann-Liouville fractional derivatives of order α are defined as:*

$$ {}_{a^+}\mathbb{D}_x^\alpha f(x) := \frac{1}{\Gamma(n-\alpha)} \frac{d^n}{dx^n} \int_a^x \frac{f(t)}{(x-t)^{\alpha+1-n}} dt = D^n I_{a^+}^{n-\alpha} f(x), \quad x > a, \tag{1.27} $$

$$ {}_{b^-}\mathbb{D}_x^\alpha f(x) := \frac{1}{\Gamma(n-\alpha)} \frac{d^n}{dx^n} \int_x^b \frac{f(t)}{(x-t)^{\alpha+1-n}} dt = D^n I_{b^-}^{n-\alpha} f(x), \quad x < b, $$

respectively.

In further discussion, unless mentioned otherwise, we denote ${}_{a^+}\mathbb{D}_x^\alpha f(x)$ by ${}_a\mathbb{D}_x^\alpha f(x)$ and ${}_{a^+}I_x^\alpha f(x)$ by ${}_aI_x^\alpha f(x)$, respectively. Also $\mathbb{D}^\alpha f(x)$ and $I^\alpha f(x)$, refer to ${}_{0^+}\mathbb{D}^\alpha f(x)$ and ${}_{0^+}I_x^\alpha f(x)$, respectively.

Proof 1.1 *For any $f \in C[a,b]$ the R-L fractional integral satisfies*

$$ {}_aI_x^\alpha \, {}_aI_x^\beta f(x) = {}_aI_x^\beta \, {}_aI_x^\alpha f(x) = {}_aI_x^{\alpha+\beta} f(x). \tag{1.28} $$

Proof.

$$ {}_aI_x^\alpha \, {}_aI_x^\beta f(x) = \frac{1}{\Gamma(\alpha)\Gamma(\beta)} \int_a^x \frac{dt}{(x-a)^{1-\alpha}} \int_a^t \frac{f(u)du}{(t-u)^{1-\beta}}, \tag{1.29} $$

and since $f \in C[a,b]$, by Fubini's theorem, the order of integration can be interchanged and by setting $t = u + s(x-u)$ we obtain:

$$ {}_aI_x^\alpha \, {}_aI_x^\beta f(x) = \frac{B(\alpha,\beta)}{\Gamma(\alpha)\Gamma(\beta)} \int_a^x \frac{f(u)du}{(x-u)^{1-\alpha-\beta}} = {}_aI_x^{\alpha+\beta} f(x), \tag{1.30} $$

Proof 1.2

$$ {}_a\mathbb{D}_x^\alpha {}_aI_x^\alpha f(x) = f(x). \tag{1.31} $$

Proof. For proving this property, we use the identity

$$ D_a^n \, I_a^n f(x) = f(x), \quad n \in \mathbb{N}. \tag{1.32} $$

Property (1.2) can be proved by induction and using Leibniz rule as follows:

$$ {}_a\mathbb{D}_x^\alpha \, {}_aI_x^\alpha f(x) = D^n \, {}_aI_x^{n-\alpha} \, {}_aI_x^\alpha f(x) = D^n I^n f(x) = f(x), \tag{1.33} $$

Proof 1.3

$$
{}_aI_x^\alpha\,{}_a\mathbb{D}_x^\alpha f(x) = f(x) - \sum_{k=0}^{n-1} [\mathbb{D}^{\alpha-k} f(x)]_{x=a} \frac{(x-a)^{\alpha-k}}{\Gamma(\alpha-k+1)}. \tag{1.34}
$$

Proof.

$$
\begin{aligned}
{}_aI_x^\alpha\,{}_a\mathbb{D}_x^\alpha f(x) &= \frac{1}{\Gamma(\alpha)} \int_a^x (x-t)^{\alpha-1} \mathbb{D}^\alpha f(t)dt \\
&= D\left\{ \frac{1}{\Gamma(\alpha+1)} \int_a^x (x-t)^\alpha \mathbb{D}^\alpha f(t)dt \right\}.
\end{aligned} \tag{1.35}
$$

By repeatedly integrating by parts and using properties (1.2), we obtain:

$$
\begin{aligned}
&\frac{1}{\Gamma(\alpha+1)} \int_a^x (x-t)^\alpha D^n \{I^{n-\alpha} f(t)\} dt \\
&= \frac{1}{\Gamma(\alpha-n+1)} \int_a^x (x-t)^{\alpha-n} I^{n-\alpha} f(t)dt \\
&\quad - \sum_{k=1}^{n} \left[D^{n-k}(I^{n-\alpha} f(x)\right]_{x=a} \frac{(x-a)^{\alpha-k+1}}{\Gamma(\alpha-k+2)} \\
&= \frac{1}{\Gamma(\alpha-n+1)} \int_a^x (x-t)^{\alpha-n} I^{n-\alpha} f(t)dt \\
&\quad - \sum_{k=1}^{n} \left[\mathbb{D}^{\alpha-k} f(x)\right]_{x=a} \frac{(x-a)^{\alpha-k+1}}{\Gamma(\alpha-k+2)} \\
&= I^{\alpha-n+1} I^{n-\alpha-1} f(x) - \sum_{k=1}^{n} \left[\mathbb{D}^{\alpha-k} f(x)\right]_{x=a} \frac{(x-a)^{\alpha-k+1}}{\Gamma(\alpha-k+2)} \\
&= If(x) - \sum_{k=1}^{n} \left[\mathbb{D}^{\alpha-k} f(x)\right]_{x=a} \frac{(x-a)^{\alpha-k+1}}{\Gamma(\alpha-k+2)}.
\end{aligned} \tag{1.36}
$$

Combining (1.35) and (1.36) ends the proof of the relationship (1.34). In particular, if $\left[\mathbb{D}^{\alpha-k} f(x)\right]_{x=a} = 0$ for $k = 0, 1, \ldots, n-1$, we have ${}_aI_x^\alpha\,{}_a\mathbb{D}_x^\alpha f(x) = f(x)$. Below we present the fractional integrals of some basic functions (Table 1.1).

TABLE 1.1 Fractional Integrals of Some Basic Functions

$f(x)$	${}_aI_x^\alpha f(x)$	**Specifications**
C	$\frac{C}{\Gamma(\alpha+1)}(x-a)^\alpha$	$\alpha \in \mathbb{R},\ a \in \mathbb{R}$
$(x-a)^\beta$	$\frac{\Gamma(\beta+1)}{\Gamma(\alpha+\beta+1)}(x-a)^{\alpha+\beta}$	$\alpha > 0,\ Re(\beta) > -1$
e^{bx}	$b^{-\alpha} e^{bx}$	$a = -\infty,\ \alpha > 0,\ Re(b) > 0$

1.3.4 Caputo Fractional Derivative

Although the R-L definition of fractional derivatives seems to play an important role in the development of fractional calculus, several authors including Caputo (1967, 1969) realized that the R-L definition needs revision because the applied problems in viscoelaticity, solid mechanics and in rheology require physically interpretable initial conditions such as $f(0)$, $f'(0)$, $f''(0)$.

Definition 1.6 *Let $f \in C^n[a,b]$ and $n-1<\alpha<n$ then*

$$ {}_aD_x^\alpha f(x) = I^{n-\alpha}D^n f(x) = \frac{1}{\Gamma(\alpha-n)}\int_a^x \frac{f^{(n)}(t)}{(x-t)^{\alpha-n+1}}dt, \quad a<x<b, \tag{1.37}$$

is called as Caputo fractional derivative.

Properties

$$(i)\ {}_aD_x^\alpha C = 0, \quad C \text{ is a constant.} \tag{1.38}$$

$$(ii)\ \lim_{\alpha\to n} {}_aD_x^\alpha f(x) = f^{(n)}(x). \tag{1.39}$$

Proof.(i) trivial.

$$\begin{aligned}(ii)\ \lim_{\alpha\to n} {}_aD_x^\alpha f(x) &= \lim_{\alpha\to n}\left(\frac{f^{(n)}(a)(x-a)^{n-\alpha}}{\Gamma(n-\alpha+1)}\right.\\ &\quad \left.+\frac{1}{\Gamma(n-\alpha+1)}\int_a^x (x-t)^{n-\alpha}f^{(n+1)}(t)dt\right)\\ &= f^{(n)}(a)+\int_a^x f^{(n+1)}(t)dt = f^{(n)}(x), \quad n=1,2,\ldots.\end{aligned}$$

$$ {}_aD_x^\alpha D^m f(x) = {}_aD_x^{\alpha+m} f(x), \quad n-1<\alpha<n, \ m\in\mathbb{N}. \tag{1.40}$$

Thus, for $\alpha \longrightarrow n$, the Caputo fractional derivative becomes the ordinary nth order derivative of the function.

1.3.4.1 Relation between Riemann-Liouville and Caputo Fractional Derivatives

Theorem 1.1 *Let $f \in C^n[a,b]$ and $n-1<\alpha<n$. Then the R-L and the Caputo fractional derivatives are connected by the relation*

$$ {}_a\mathbb{D}_x^\alpha f(x) = {}_aD_x^\alpha f(x) + \sum_{k=0}^{n-1}\frac{f^{(k)}(a^+)}{\Gamma(1+k-\alpha)}(x-a)^{k-\alpha}. \tag{1.41}$$

Proof.

$$ {}_a\mathbb{D}_x^\alpha f(x) = D^n\, {}_aI_x^{n-\alpha} f(x) = D^n\left[{}_aI_x^{n-\alpha}\left(I^n f^{(n)}(x) + \sum_{k=0}^{n-1}\frac{f^{(k)}(a^+)}{k!}(x-a)^k\right)\right]$$

$$= {}_aI_x^{n-\alpha} f^{(n)}(x) + D^n {}_aI_x^{n-\alpha} \sum_{k=0}^{n-1} \frac{f^{(k)}(a^+)}{k!}(x-a)^k \tag{1.42}$$

$$= {}_aD_x^{\alpha} f(x) + \sum_{k=0}^{n-1} \frac{f^{(k)}(a^+)}{\Gamma(1+k-\alpha)}(x-a)^{k-\alpha}. \tag{1.43}$$

From the above theorem, we get the following result:

i. If $\alpha = n \in \mathbb{N}$, then ${}_a\mathbb{D}_x^{\alpha} f(x) = {}_aD_x^{\alpha} f(x) = D^n f(x)$.

ii. If $f^{(k)}(a) = 0$ for $k = 0, 1, \ldots, n-1$, then $\mathbb{D}_a^{\alpha} f(x) = D_a^{\alpha} f(x)$.

iii. If $0 < \alpha < 1$, then ${}_a\mathbb{D}_x^{\alpha} f(x) = {}_aD_x^{\alpha} f(x) + \dfrac{f(a)}{\Gamma(1-\alpha)}(x-a)^{-\alpha}$.

Theorem 1.2 *Let $f \in C^n[a,b]$ and $n-1 < \alpha < n$ then*

$${}_aI_x^{\alpha}\, {}_aD_x^{\alpha} f(x) = f(x) - \sum_{k=0}^{n-1} \frac{f^{(k)}(a^+)}{k!}(x-a)^k, \; x \geq a. \tag{1.44}$$

Proof.

$${}_aI_x^{\alpha}\, {}_aD_x^{\alpha} f(x) = I_a^{\alpha}\, {}_aI_x^{n-\alpha} f^{(n)}(x) = I^{(n)} f^{(n)}(x) = f(x) - \sum_{k=0}^{n-1} \frac{f^{(k)}(a)}{k!}(x-a)^k, \; x \geq a.$$

Equation (1.44) is a particular case of the more general property

$$\begin{aligned} {}_aI_x^{\alpha}\, {}_aD_x^{\beta} f(x) &= {}_aI_x^{\alpha}\, {}_aI_x^{m-\alpha} f^{(m)}(x) \\ &= {}_aI_x^{\alpha-\beta}\left(I^{(n)} f^{(n)}(x)\right) \qquad \alpha > \beta, \\ &= {}_aI_x^{\alpha-\beta} f(x) - \sum_{k=0}^{n-1} \frac{f^{(k)}(a)}{\Gamma(\alpha-\beta+k+1)}(x-a)^{\alpha-\beta+k}, \\ & \quad x \geq a, \; m-1 < \beta < m. \end{aligned} \tag{1.45}$$

Different types of fractional derivatives have been presented during the last few decades such as Caputo-Fabrizio, Atangana-Baleanu, conformable fractional derivatives and so on. In the rest of this book, we just focused on fractional differential equations involving R-L and Caputo fractional derivatives.

CHAPTER 2

Decomposition Methods for Fractional Differential Equations

VARIOUS METHODS, for example, Laplace and Fourier transforms, have been utilized to solve linear fractional differential equations (FDE). In contrast for solving the nonlinear FDEs, one has to depend upon numerical solutions solely. Numerical methods use discretization that gives rise to rounding off errors causing loss of accuracy and requires large computer power and time.

So, in this chapter, we apply the Adomian decomposition, Homotopy perturbation, Homotopy analysis and Daftardar-Jafari iterative methods for solving fractional differential equations. Also, we apply these methods for solving system of linear and nonlinear fractional differential equations. A few modifications of these methods are presented.

2.1 ADOMIAN DECOMPOSITION METHOD

The Adomian decomposition method (ADM) [2] has proven to be a powerful method and has successfully been applied in a variety of problems. Since the beginning of the 1980s, the Adomian decomposition method has been applied to a wide class of functional equations [17, 18, 57, 58]. In this method, the solution is given as an infinite series usually converging to an accurate solution.

2.1.1 Systems of Linear Fractional Differential Equations

In this section, we consider the following system of linear fractional differential equations [18, 49]

$$D^{\alpha_i} y_i(t) = \sum_{j=1}^{n} (\phi_{ij}(t) + \gamma_{ij} D^{\alpha_{ij}}) y_j + g_i(t), \tag{2.1}$$

DOI: 10.1201/9781003474609-2

with the initial conditions

$$y_i^{(k)}(0) = c_k^i, \tag{2.2}$$

where $0 \leq k \leq [\alpha_i]$, if α_i is not an integer, and $0 \leq k \leq \alpha_i - 1$, if α_i is an integer. Here $0 \leq \alpha_{ij} < \alpha_i$, for $1 \leq i, j \leq n, \gamma_{ij}$'s are constants and $\phi_{ij}(t)$, $g_i(t) \in C[0,T]$. Applying I^{α_i} to both the sides of Eq. (2.1), we get

$$\begin{aligned} y_i(t) &= \sum_{k=0}^{[\alpha_i]} c_k^i \frac{t^k}{k!} + I^{\alpha_i} g_i(t) - \sum_{j=1}^{n} \gamma_{ij} \sum_{k=0}^{[\alpha_{ij}]} c_k^j \frac{t^{\alpha_i - \alpha_{ij} + k}}{\Gamma(\alpha_i - \alpha_{ij} + k + 1)} \\ &+ \sum_{j=1}^{n} (I^{\alpha_i} \phi_{ij}(t) + \gamma_{ij} I^{\alpha_i - \alpha_{ij}}) y_j, \text{ where } 1 \leq i \leq n. \end{aligned} \tag{2.3}$$

We employ Adomian decomposition method to solve the system of equations (2.3). The Adomian decomposition method [2] consists of representing y_i in the decomposition form given by

$$y_i(t) = \sum_{m=0}^{\infty} y_{im}(t), \tag{2.4}$$

where the components y_{im}, $m \geq 0$, can be determined in a recursive manner. Substituting Eq. (2.4) into both sides of Eq. (2.3), we get

$$\begin{aligned} \sum_{m=0}^{\infty} y_{im}(t) &= \sum_{k=0}^{[\alpha_i]} c_k^i \frac{t^k}{k!} + I^{\alpha_i} g_i(t) - \sum_{j=1}^{n} \gamma_{ij} \sum_{k=0}^{[\alpha_{ij}]} c_k^j \frac{t^{\alpha_i - \alpha_{ij} + k}}{\Gamma(\alpha_i - \alpha_{ij} + k + 1)} \\ &+ \sum_{m=0}^{\infty} (\sum_{j=1}^{n} (I^{\alpha_i} \phi_{ij}(t) + \gamma_{ij} I^{\alpha_i - \alpha_{ij}}) y_{jm}(t)), \; 1 \leq i \leq n. \end{aligned} \tag{2.5}$$

The decomposition method defines the components $y_{im}(x)$, $m \geq 0$, by the following recursion relation.

$$\begin{aligned} y_{i0}(t) &= \sum_{k=0}^{[\alpha_i]} c_k^i \frac{t^k}{k!} + I^{\alpha_i} g_i(t) - \sum_{j=1}^{n} \gamma_{ij} \sum_{k=0}^{[\alpha_{ij}]} c_k^j \frac{t^{\alpha_i - \alpha_{ij} + k}}{\Gamma(\alpha_i - \alpha_{ij} + k + 1)}, \\ y_{i,m+1}(t) &= \sum_{j=1}^{n} (I^{\alpha_i} \phi_{ij}(t) + \gamma_{ij} I^{\alpha_i - \alpha_{ij}}) y_{jm}(t), \; 1 \leq i \leq n, \; m = 0, 1, \ldots \end{aligned} \tag{2.6}$$

We approximate the solution $y_i(x)$ by the truncated series:

$$y_{ik}(t) = \sum_{m=0}^{k-1} y_{im}(t), \quad \text{and} \quad y_i(t) = \lim_{k \to \infty} y_{ik}(t).$$

Adomian decomposition method is very simple in its principles, though the difficulties consist in proving the convergence of the Adomian series. In the following section, we prove the convergence of the series $\sum_{m=0}^{\infty} y_{im}(t)$.

2.1.1.1 Convergence of Adomian Decomposition Method

In this section, we show convergence of the decomposition series [18, 49].

In view of Eq. (2.6)

$$\begin{aligned} |y_{i1}(t)| &\leq \sum_{j=1}^{n} |I^{\alpha_i}\phi_{ij}(t)y_{j0}(t)| + \left|\gamma_{ij}I^{\alpha_i-\alpha_{ij}}y_{j0}(t)\right| \\ &\leq nML\frac{t^{\alpha_i}}{\Gamma(\alpha_i+1)} + \gamma L\sum_{j=1}^{n}\frac{t^{\alpha_i-\alpha_{ij}}}{\Gamma(\alpha_i-\alpha_{ij}+1)}, \end{aligned}$$

where $M = \max_{t\in[0,T]}\{\phi_{ij}(t)\}, i,j = 1,\dots n, \gamma = max_{1\leq i,j\leq n}\{\gamma_{ij}\}$, $L = \max_{x\in[0,T]} y_{i0}(t)$.

Let $\beta_{ij} = \alpha_i - \alpha_{ij}, 1 \leq j \leq n$, $\beta_{i,n+1} = \alpha_i$, $\delta > max\{nML, \gamma L\}$ and $z_{ij} = \delta t^{\beta_{ij}}, 1 \leq i,\, j \leq n$.

$$|y_{i1}(t)| \leq \delta\sum_{j=1}^{n+1}\frac{t^{\beta_{ij}}}{\Gamma(1+\beta_{ij})} = \sum_{\substack{l_1+\dots+l_{n+1}=1\\ l_i\geq 0}} (k; l_1,\dots,l_{n+1})\left[\frac{\prod_{j=1}^{n+1} z_{ij}^{l_i}}{\Gamma(\beta+\sum_{j=1}^{n+1}\beta_{ij}l_j)}\right],$$

$$\begin{aligned} |y_{i2}(t)| &\leq \delta^2\sum_{j=1}^{n+1} I^{\beta_{ij}}\left[\sum_{k=1}^{n+1}\frac{t^{\beta_{ik}}}{\Gamma(1+\beta_{ik})}\right] \\ &= \delta^2\sum_{j,k=1}^{n+1}\frac{t^{\beta_{ij}+\beta_{ik}}}{\Gamma(1+\beta_{ij}+\beta_{ik})} \\ &= \sum_{\substack{l_1+\dots+l_{n+1}=2\\ l_i\geq 0}} (k; l_1,\dots,l_{n+1})\left[\frac{\prod_{j=1}^{n+1} z_{ij}^{l_i}}{\Gamma(1+\sum_{j=1}^{n+1}\beta_{ij}l_j)}\right], \\ &\vdots \end{aligned}$$

$$
\begin{aligned}
|y_{im}(t)| &\leq \delta^m \sum_{j_1=1}^{n+1} \sum_{j_2=1}^{n+1} \cdots \sum_{j_m=1}^{n+1} \frac{\prod_{k=1}^{m} t^{\beta_{ij_k}}}{\Gamma(1+\sum_{k=1}^{m} \beta_{ij_k})} \\
&= \sum_{\substack{l_1+\cdots+l_{n+1}=m \\ l_i \geq 0}} (k; l_1, \ldots, l_{n+1}) \left[\frac{\prod_{j=1}^{n+1} z_{ij}^{l_i}}{\Gamma(1+\sum_{j=1}^{n+1} \beta_{ij} l_j)} \right], \\
&\ \vdots
\end{aligned}
$$

Hence $\sum_{m=0}^{\infty} y_{im} \leq \mathcal{E}^i_{(\beta_{i1},\ldots,\beta_{i,n+1}),\beta=1}(z_{i1}, z_{i2}, \ldots, z_{i,n+1})$, where $\beta = 1$ and $z_{ij} = \delta t^{\beta_{ij}}$, $j = 1, \ldots, n+1$. The multivariate Mittag-Leffler function $\mathcal{E}^i$ [92] is defined below.

$$
\mathcal{E}^i_{(\beta_{i1},\ldots,\beta_{i,n+1}),\beta}(z_{i1}, z_{i2}, \ldots, z_{i,n+1}) = \sum_{k=0}^{\infty} \sum_{\substack{l_1+\cdots+l_{n+1}=k \\ l_i \geq 0}} (k; l_1, \ldots, l_{n+1}) \times \left[\frac{\prod_{j=1}^{n+1} z_{ij}^{l_i}}{\Gamma(\beta+\sum_{j=1}^{n+1} \beta_{ij} l_j)} \right].
$$

Thus $|y_{im}(t)|$ is bounded by the mth term of a multivariate Mittag-Leffler series, which is convergent. Hence by comparison test $\sum_{m=0}^{\infty} y_{im}(t)$ is convergent.

2.1.1.2 Applications

Example 2.1 *Consider the initial value problem [18, 49]:*

$$
D^{\bar{\alpha}} \bar{\boldsymbol{y}} = \boldsymbol{A} \bar{\boldsymbol{y}}, \quad \bar{y}(0) = (c_1, \ldots, c_n)^T, \qquad 0 < \alpha_i < 1, \quad i = 1, \ldots, n,
$$

where $\bar{\boldsymbol{y}} = (y_1, \ldots, y_n)^T$, $D^{\bar{\alpha}} \bar{\boldsymbol{y}} = (D^{\alpha_1} y_1, \ldots, D^{\alpha_n} y_n)^T$ *and* $\boldsymbol{A} = [a_{ij}]_{n \times n}$ *is a square matrix of constants.*

This system is equivalent to the following system of integral equations

$$
y_i(t) = c_i + I^{\alpha_i} \sum_{j=1}^{n} a_{ij} y_j(t), \quad i = 1, 2, \ldots, n.
$$

Note that this is a special case of Eq. (2.1) and is obtained by putting $\phi_{ij} = 0 = g_i$. In view of Eq. (2.6):

$$\begin{aligned}
y_{i0} &= c_i, \quad y_{i1} = \sum_{j=1}^{n} a_{ij}\frac{t^{\alpha_i}}{\Gamma(1+\alpha_i)}, \\
y_{i2} &= \sum_{j,k=1}^{n} a_{ij}a_{jk}c_k\frac{t^{\alpha_i+\alpha_j}}{\Gamma(1+\alpha_i+\alpha_j)}, \\
y_{i3} &= \sum_{j,k,s=1}^{n} a_{is}a_{ij}a_{jk}c_k\frac{t^{\alpha_i+\alpha_s+\alpha_j}}{\Gamma(1+\alpha_i+\alpha_j)}, \\
&\vdots \\
y_{im} &= \sum_{l_1,l_2,\ldots,l_m=1}^{n} a_{l_1l_2}a_{l_1l_2}\cdots a_{l_{m-1}l_m}c_{l_m}\left[\frac{t^{\sum_{j=1}^{m}\alpha_{l_j}}}{\Gamma(1+\sum_{j=1}^{m}\alpha_{l_j})}\right].
\end{aligned}$$

It is observed that $\bar{y}(x) = [\mathcal{E}_{(\alpha_1,\ldots,\alpha_n),1}(t^{\alpha_1}A_1,\ldots,t^{\alpha_n}A_n)]\bar{y}(0)$ where A_i has the only non-zero row as i-th row with entries $[a_{i1},\ldots a_{in}]$, and all other entries are zero. Here $\mathcal{E}_{(\alpha_1,\ldots,\alpha_n),1}(t^{\alpha_1}A_1,\ldots,t^{\alpha_n}A_n)$ denotes multivariate Mittag-Leffler function for matrix arguments $A_1,\ldots,A_n$ as follows:

$$\begin{aligned}
\mathcal{E}_{(\alpha_1,\ldots,\alpha_n),\beta}(A_1,\ldots,A_n) &= I + \sum_{i_1=1}^{n}\frac{A_{i_1}}{\Gamma(\beta+\alpha_{i_1})} \\
&+ \sum_{i_1=1,i_2=1}^{n}\frac{A_{i_1}A_{i_2}}{\Gamma(\beta+\alpha_{i_1}+\alpha_{i_2})} + \cdots \\
&+ \sum_{i_1=1,i_2=1,\ldots,i_m=1}^{n}\frac{A_{i_1}A_{i_2}\ldots A_{i_m}}{\Gamma(\beta+\alpha_{i_1}+\alpha_{i_2}+\cdots+\alpha_{i_m})} + \cdots \\
&= I + \sum_{m=1}^{\infty}\left[\sum_{i_1=1,i_2=1,\ldots,i_m=1}^{n}\frac{A_{i_1}A_{i_2}\ldots A_{i_m}}{\Gamma(\beta+\alpha_{i_1}+\alpha_{i_2}+\cdots+\alpha_{i_m})}\right].
\end{aligned} \tag{2.7}$$

Example 2.2 *Consider the fractional oscillation equation [18]*

$$D^{1+\alpha}y(t) + by(t) = f(t),$$

with initial conditions

$$y(0) = y_0, \quad y'(0) = y'_0, \tag{2.8}$$

where $\alpha \in (0,1)$ and f satisfies Lipschitz condition.

This equation is equivalent to the following system of equations:

$$D^1 y_1 = y_2, \quad y_1(0) = y_0,$$
$$D^\alpha y_2 = -by_1 + f(t), \quad y_2(0) = y_0'.$$

Here we solve the following particular case of the above system ($b = 1$ and $f(t) = 0$) using Adomian decomposition method.

$$\begin{aligned} D^1 y_1 &= y_2, \quad y_1(0) = 1, \\ D^\alpha y_2 &= -y_1, \quad y_2(0) = 0. \end{aligned} \tag{2.9}$$

To derive the solution, we use the following Adomian scheme,

$$\begin{aligned} y_1 &= 1 + I^1 y_2, \\ y_2 &= -I^\alpha y_1, \\ y_{10} &= 1, \\ y_{20} &= 0, \\ y_{1,m+1} &= I y_{2m}, \\ y_{2,m+1} &= -I^\alpha y_{1m}, \quad m = 0, 1, \ldots. \end{aligned}$$

In the first iteration, we have:

$$\begin{aligned} y_{11} &= I y_{20} = 0, \\ y_{21} &= -I^\alpha y_{10} = -\frac{\Gamma(1)}{\Gamma(\alpha+1)} t^\alpha. \end{aligned}$$

The subsequent terms are

$$\begin{aligned} y_{12} &= I y_{21} = -\frac{t^{\alpha+1}}{\Gamma(\alpha+2)}, \\ y_{22} &= -I^\alpha y_{11} = 0, \\ y_{13} &= I y_{22} = 0, \\ y_{23} &= -I^\alpha y_{12} = -\frac{t^{2\alpha+1}}{\Gamma(2\alpha+2)}. \end{aligned}$$

In general, we get

$$y_{1j} = \begin{cases} (-1)^k \frac{t^{k(\alpha+1)}}{\Gamma(k(\alpha+1)+1)}, & j = 2k, \\ 0, & j = 2k+1, \end{cases} \tag{2.10}$$

$$y_{2j} = \begin{cases} 0, & j = 2k, \\ (-1)^{(k+1)} \frac{t^{k(\alpha+1)+\alpha}}{\Gamma[(\alpha+1)(k+1)]}, & j = 2k+1. \end{cases} \tag{2.11}$$

Using Eq. (2.4), we can write y_1 and y_2 in the following form:

$$y_1 = \sum_{j=0}^{\infty} y_{1j} = \sum_{k=0}^{\infty} (-1)^k \frac{t^{k(\alpha+1)}}{\Gamma(k(\alpha+1)+1)},$$

$$y_2 = \sum_{j=0}^{\infty} y_{2j} = \sum_{k=0}^{\infty}(-1)^{(k+1)}\frac{t^{k(\alpha+1)+\alpha}}{\Gamma[(\alpha+1)(k+1)]}. \tag{2.12}$$

The approximation solutions of y_1 for various values of α using ADM are obtained. The numerical results are shown in Figures 2.1 and 2.2, respectively.

Example 2.3 *We discuss Bagley–Torvik equation that arises, for instance, in modeling the motion of a rigid plate immersed in Newtonian fluid,*

$$D^2y + b_2D^{\alpha_1}y + b_1y = f, \quad y(0) = y_0, \quad y'(0) = y_0',$$

where $\alpha_1 = \alpha + 1$, $\alpha \in (0, 1)$.

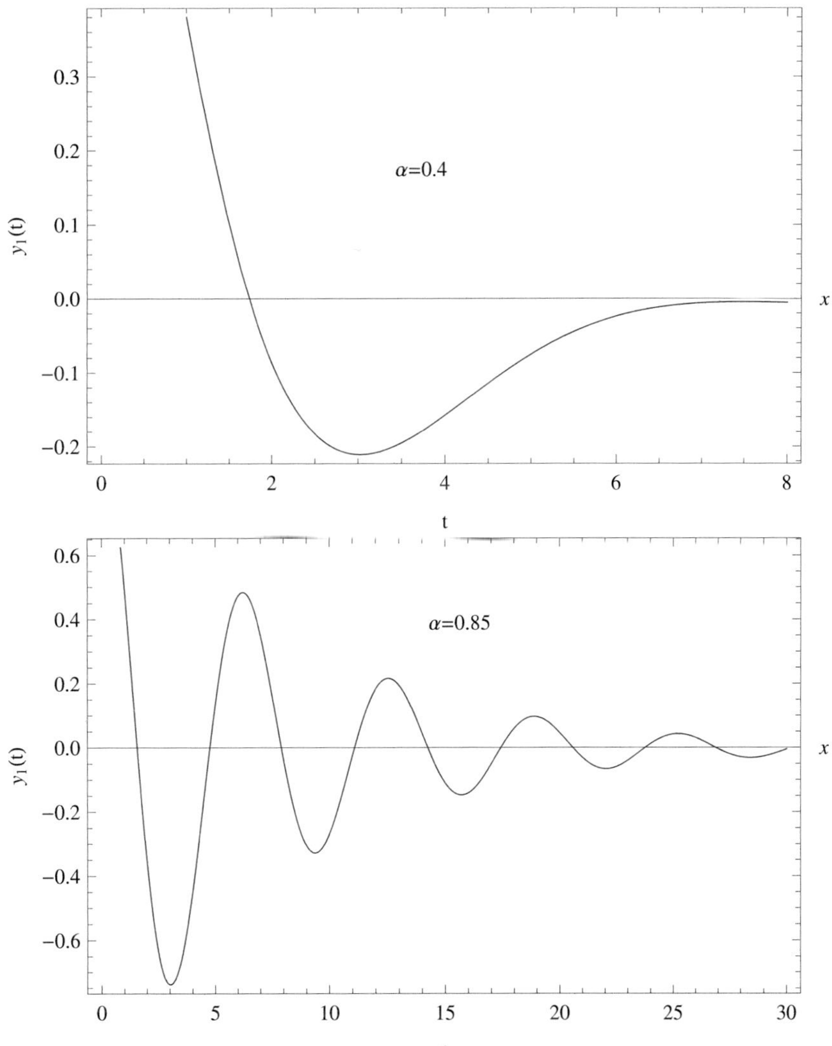

Figure 2.1 Approximate solutions $y_1(t)$ using ADM for Example 2.2.

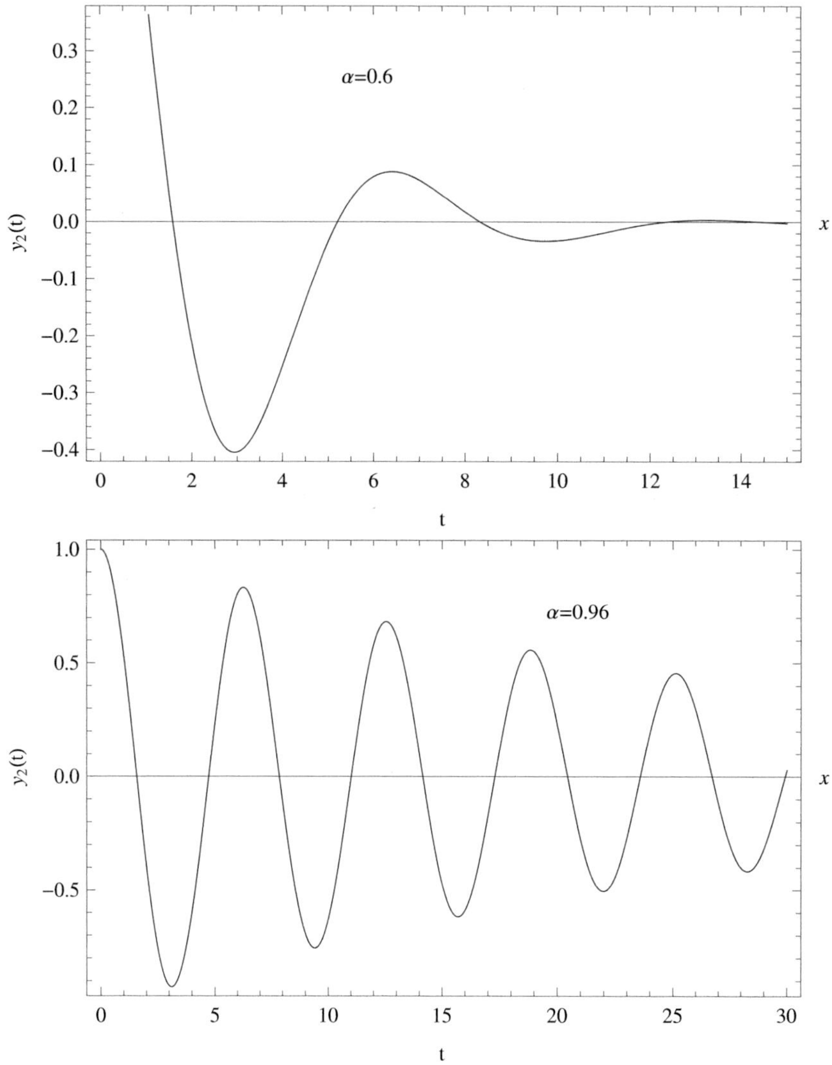

Figure 2.2 Approximate solutions $y_2(t)$ using ADM for Example 2.2.

This equation can be viewed as the following system of equations [18]

$$\begin{aligned} &D^{\beta} y_1 = y_2, \quad y_1(0) = y_0, \\ &D^{\beta} + b_2 D^{\alpha} y_2 = f - b_1 y_1, \quad y_2(0) = y'(0), \end{aligned}$$

where $\beta = 1$. Using the Adomian decomposition method scheme:

$$y_1 = \sum_{k=0}^{[\beta]} c_k^1 \frac{t^k}{k!} - I^{\beta} y_2,$$

$$y_2 = \sum_{k=0}^{[\beta]} c_k^2 \frac{t^k}{k!} - I^{\beta-\alpha} y_2 + \sum_{k=0}^{[\alpha]} c_k^2 \frac{t^{\beta-\alpha+k}}{\Gamma(\beta - \alpha + k + 1)} - I^{\beta} y_1,$$

$$\begin{cases} y_{10} = 1, \\ y_{1,m+1} = I^{\beta} y_{2m}. \end{cases} \quad m = 0, 1, 2, \ldots, \tag{2.13}$$

$$\begin{cases} y_{20} = 0, \\ y_{2,m+1} = -I^{\beta-\alpha} y_{2m} - I^{\beta} y_{1m}, \end{cases} \quad m = 0, 1, 2, \ldots. \tag{2.14}$$

In the first iteration, we get

$$\begin{cases} y_{11} = I^{\beta} y_{20} = 0, \\ y_{21} = -I^{\beta-\alpha} y_{20} - I^{\beta} y_{10} = -\frac{t^{\alpha}}{\Gamma(\alpha+1)}, \end{cases}$$

and the following terms are

$$\begin{cases} y_{12} = I^{\beta} y_{21} = -\frac{t^{\alpha+\beta}}{\Gamma(\alpha+\beta+1)}, \\ y_{22} = \frac{t^{2\beta-\alpha}}{\Gamma(2\beta-\alpha+1)}, \end{cases}$$

$$\begin{cases} y_{13} = \frac{t^{3\beta-\alpha}}{\Gamma(3\beta-\alpha+1)}, \\ y_{23} = -\frac{t^{3\beta-2\alpha}}{\Gamma(3\beta-2\alpha+1)} + \frac{t^{3\beta}}{\Gamma(3\beta+1)}, \end{cases}$$

$$\begin{cases} y_{14} = \frac{t^{4\beta-2\alpha}}{\Gamma(4\beta-2\alpha+1)} + \frac{t^{4\beta}}{\Gamma(4\beta+1)}, \\ y_{24} = \frac{t^{4\beta-3\alpha}}{\Gamma(4\beta-3\alpha+1)} - 2\frac{t^{4\beta-\alpha}}{\Gamma(4\beta-\alpha+1)}, \end{cases}$$

$$y_{1n} = \sum_{j=1}^{[\frac{n}{2}]} (-1)^{n+j} a_{nj} \frac{t^{n(\beta-\alpha)+2j\alpha}}{\Gamma(n(\beta-\alpha)+2j\alpha+1)}, \tag{2.15}$$

where

$$a_{nj} = \begin{cases} a_{n-1,j} + a_{n-2,j-1}, & 1 \le j \le \frac{n}{2}, \\ 1, & n = j = 0, \\ 0, & \text{otherwise}. \end{cases} \tag{2.16}$$

Following Eq. (2.4) we can write y_1 and y_2 in the following form

$$\begin{aligned} y_1 = \sum_{n=0}^{\infty} y_{1n} &= 1 + \sum_{n=0}^{\infty} \sum_{j=1}^{[\frac{n}{2}]} (-1)^{n+j} a_{nj} \frac{t^{n(\beta-\alpha)+2j\alpha}}{\Gamma(n(\beta-\alpha)+2j\alpha+1)}, \\ y_2 = \sum_{n=0}^{\infty} y_{2n} &= \sum_{n=2}^{\infty} \sum_{j=1}^{[\frac{n}{2}]} (-1)^{n+j+1} b_{nj} \frac{t^{n(\beta-\alpha)+(2j-1)\alpha}}{\Gamma(n(\beta-\alpha)+(2j-1)\alpha+1)} \\ &+ \sum_{k=0}^{[\frac{n}{2}]} (-1)^{k} \frac{t^{2k}}{\Gamma(2k+1)}, \end{aligned}$$

where b_{nj} is defined similar to a_{nj}.

Figure 2.3 displays approximate solutions of y_1 for $\alpha = 0.5$ $\beta = 1$ and $\alpha = 0.25$, $\beta = 1$.

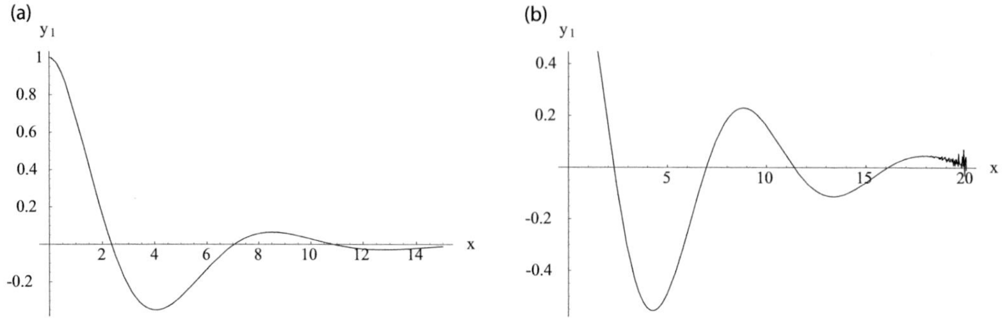

Figure 2.3 Approximate solutions $y_1(x)$ using ADM $\alpha = 0.25$ (a) and $\alpha = 0.5$ (b) for Example 2.3.

Remark 2.1 *The graphs drawn in the Figure 2.3 are in excellent agreement with those drawn in Ref. [26] using numerical methods.*

2.1.2 Systems of Nonlinear Fractional Differential Equations

In this section, Adomian method has been applied to the following system of nonlinear fractional differential equations as Ref. [57]:

$$D^{\alpha_i} y_i(t) = N_i(t, y_1, \dots, y_n), \qquad y_i^{(k)}(0) = c_k^i, \quad 0 \le k \le [\alpha_i], \tag{2.17}$$

where $1 \le i \le n$, and $\alpha_i \in \Re^+$ and N_i's are linear/nonlinear functions of t, y_1, $\dots$, y_n.

Applying I^{α_i} to both the sides of Eq. (2.17), we get

$$y_i = \sum_{k=0}^{[\alpha_i]} c_k^i \frac{t^k}{k!} + I^{\alpha_i} N_i(t, y_1, \dots, y_n), \;\; 1 \le i \le n. \tag{2.18}$$

We employ Adomian decomposition method to solve the system of equations (2.17). Let

$$y_i = \sum_{m=0}^{\infty} y_{im}, \tag{2.19}$$

and

$$N_i(t, y_1, \dots, y_n) = \sum_{m=0}^{\infty} A_{im}, \tag{2.20}$$

where A_{im} are Adomian polynomials which depend upon $y_{10}, \dots, y_{1m}, y_{20}, \dots, y_{2m}, \dots, y_{n0}, \dots, y_{nm}$. In view of Eqs. (2.19) and (2.20), (2.18) takes the form:

$$\sum_{m=0}^{\infty} y_{im} = \sum_{k=0}^{[\alpha_i]} c_k^i \frac{t^k}{k!}$$
$$+ I^{\alpha_i} \sum_{m=0}^{\infty} A_{im}(y_{10}, \dots, y_{1m}, \dots, y_{n0}, \dots, y_{nm}), \qquad 1 \le i \le n. \quad (2.21)$$

We set

$$y_{i0}(t) = \sum_{k=0}^{[\alpha_i]} c_k^i \frac{t^k}{k!}, \quad (2.22)$$
$$y_{i,m+1}(t) = I^{\alpha_i} A_{im}(y_{10}, \dots, y_{1m}, \dots, y_{n0}, \dots, y_{nm}),\ 1 \le i \le n,\ m = 0, 1, \dots.$$

In order to determine the Adomian polynomials, we introduce a parameter λ and (2.20) becomes

$$N_i(t, \sum_{m=0}^{\infty} y_{1m}\lambda^m, \dots, \sum_{m=0}^{\infty} y_{nm}\lambda^m) = \sum_{m=0}^{\infty} A_{im}\lambda^m. \quad (2.23)$$

Let $y_{i\lambda}(t) = \sum_{m=0}^{\infty} y_{im}(t)\lambda^m$, then

$$A_{im} = \frac{1}{m!}\left[\frac{d^m}{d\lambda^m} N_{i\lambda}(y_1, \dots, y_n)\right]_{\lambda=0}, \quad (2.24)$$

where

$$N_{i\lambda}(y_1, \dots, y_n) = N_i(t, y_{1\lambda}, \dots, y_{n\lambda}). \quad (2.25)$$

In view of Eqs. (2.24) and (2.25), we get

$$\begin{aligned} A_{im} &= \frac{1}{m!}\frac{d^m}{d\lambda^m}\left[N_i(t, y_{1\lambda}, \dots, y_{n\lambda})\right]_{\lambda=0} \\ &= \frac{1}{m!}\frac{d^m}{d\lambda^m}\left[N_i\left(t, \sum_{j=0}^{\infty} y_{1j}\lambda^j, \dots, \sum_{j=0}^{\infty} y_{nj}\lambda^j\right)\right]_{\lambda=0} \\ &= \left[\frac{1}{m!}\frac{d^m}{d\lambda^m} N_i\left(t, \sum_{j=0}^{m} y_{1j}\lambda^j, \dots, \sum_{j=0}^{m} y_{nj}\lambda^j\right)\right]_{\lambda=0}. \end{aligned} \quad (2.26)$$

Hence Eqs. (2.23) and (2.26) lead to the following recurrence relations.

$$\begin{aligned} y_{i0}(t) &= \sum_{k=0}^{[\alpha_i]} c_k^i \frac{t^k}{k!}, \quad (2.27) \\ y_{i,m+1}(t) &= I^{\alpha_i}\left[\frac{1}{m!}\frac{d^m}{d\lambda^m} N_i\left(t, \sum_{j=0}^{m} y_{1j}\lambda^j, \dots, \sum_{j=0}^{m} y_{nj}\lambda^j\right)\right]_{\lambda=0}, \ m = 0, 1, \dots. \end{aligned}$$

We can approximate the solution y_i by the truncated series:

$$f_{ik} = \sum_{m=0}^{k-1} y_{im}, \qquad \lim_{k\to\infty} f_{ik} = y_i(t).$$

For the convergence of the above method, we refer the reader to the work of Abboui and Cherruault [1]. If the system (2.17) admits unique solution, then this method will produce the unique solution. If the system (2.17) does not possess unique solution, the decomposition method will give a solution among many (possible) other solutions.

2.1.2.1 Applications

To demonstrate the effectiveness of the method, we consider here some systems of nonlinear fractional differential equations. Daftardar-Gejji and Babakhani [17] have presented analysis of such a system.

Example 2.4 *Consider the system of nonlinear fractional differential equations [57]*

$$\begin{cases} D^{\alpha} y_1 = y_1^2 + y_2, \\ D^{\beta} y_2 = y_2 \cos y_1, \end{cases} \qquad y_1(0) = 0, \quad y_2(0) = 1,$$

where $\alpha, \beta \in (0,1)$.

In view of the results obtained by Daftardar-Gejji and Babakhani Reference [17], this system has unique solution. In order to solve the above system, we define the nonlinear terms by

$$N_1 \bar{y} = y_1^2 + y_2 = \sum_{j=0}^{\infty} A_{1j}, \quad N_2 \bar{y} = y_2 \cos y_1 = \sum_{j=0}^{\infty} A_{2j}.$$

In view of Eq. (2.26) and using *Mathematica* software, we evaluate the Adomian polynomials. They are as follows:

$$\begin{aligned} A_{10} &= y_{10}^2 + y_{20}, \\ A_{11} &= 2y_{10}y_{11} + y_{21}, \\ A_{12} &= y_{11}^2 + 2y_{10}y_{12} + y_{22}, \\ A_{13} &= 2y_{11}y_{12} + 2y_{10}y_{13} + y_{23}, \\ A_{14} &= y_{12}^2 + 2y_{11}y_{13} + 2y_{10}y_{14} + y_{24}, \\ &\vdots \end{aligned}$$

and

$$\begin{aligned}
A_{20} &= y_{20}\cos y_{10},\\
A_{21} &= -y_{11}y_{20}\sin y_{10} + y_{21}\cos y_{10},\\
A_{22} &= \frac{1}{2}y_{11}^2 y_{20}\cos y_{10} - y_{12}y_{20}\sin y_{10} - y_{11}y_{21}\sin y_{10} + y_{21}\cos y_{10},\\
A_{23} &= \frac{1}{6}((y_{11}^3 y_{20}\sin y_{10} - 6y_{11}y_{12}\sin y_{10} - 6y_{13}\sin y_{10})y_{20}\\
&\quad -3(y_{11}^2\cos y_{10} + y_{12}\sin y_{10})y_{21} - 6y_{11}y_{22}\sin y_{10} + 6y_{23}\cos y_{10}),\\
A_{24} &= \frac{1}{24}((y_{11}^4\cos y_{10} + 12y_{11}^2 y_{12}\sin y_{10} - 24y_{11}y_{13}\cos y_{10} - 12(y_{12}^2\cos y_{10}\\
&\quad +2y_{14}\sin y_{10}))y_{20} + 4(y_{11}^3\sin y_{10} - 6y_{11}y_{12}\cos y_{10} - 6y_{13}\sin y_{10})y_{21}\\
&\quad -12(y_{11}^2\cos y_{10} + y_{12}\sin y_{10})y_{22} - 24y_{11}y_{23}\sin y_{10} + 24y_{24}\cos y_{10}),\\
&\vdots
\end{aligned}$$

The Adomian decomposition series (2.23) leads to the following scheme:

$$\begin{aligned}
y_{10} &= 0, \qquad y_{1,m+1} = I^{\alpha}A_{1m},\\
y_{20} &= 1, \qquad y_{2,m+1} = I^{\beta}A_{2m}, \quad m = 0, 1, \ldots.
\end{aligned}$$

In the first iteration, we have:

$$y_{11} = I^{\alpha}A_{10} = \frac{t^{\alpha}}{\Gamma(\alpha+1)} \text{ and } y_{21} = -I^{\beta}A_{20} = \frac{t^{\beta}}{\Gamma(\beta+1)}.$$

The subsequent terms are

$$\begin{aligned}
y_{12} &= I^{\alpha}A_{11} = I^{\alpha}\frac{t^{\beta}}{\Gamma(\beta+1)} = \frac{t^{\alpha+\beta}}{\Gamma(\alpha+\beta+1)},\\
y_{22} &= I^{\beta}A_{21} = I^{\beta}\frac{t^{\beta}}{\Gamma(\beta+1)} = \frac{t^{2\beta}}{\Gamma(2\beta+1)},\\
y_{13} &= I^{\alpha}A_{12} = I^{\alpha}\left(\frac{t^{2\alpha}}{[\Gamma(\alpha+1)]^2} + \frac{t^{2\beta}}{\Gamma(2\beta+1)}\right)\\
&= \frac{\Gamma(2\alpha+1)t^{3\alpha}}{[\Gamma(\alpha+1)]^2\Gamma(3\alpha+1)} + \frac{t^{\alpha+2\beta}}{\Gamma(\alpha+2\beta+1)},\\
y_{23} &= I^{\beta}A_{22} = I^{\beta}\left(-\frac{t^{2\alpha}}{2[\Gamma(\alpha+1)]^2} + \frac{t^{2\beta}}{\Gamma(2\beta+1)}\right)\\
&= -\frac{\Gamma(2\alpha+1)t^{2\alpha+\beta}}{[\Gamma(\alpha+1)]^2\Gamma(2\alpha+\beta+1)} + \frac{t^{3\beta}}{\Gamma(3\beta+1)}.
\end{aligned}$$

Using the above terms,

$$
\begin{aligned}
y_1 &= \frac{t^{\alpha}}{\Gamma(\alpha+1)} + \frac{t^{\alpha+\beta}}{\Gamma(\alpha+\beta+1)} + \frac{\Gamma(2\alpha+1)t^{3\alpha}}{[\Gamma(\alpha+1)]^2\Gamma(3\alpha+1)} + \frac{t^{\alpha+2\beta}}{\Gamma(\alpha+2\beta+1)} + \cdots,\\
y_2 &= 1 + \frac{t^{\beta}}{\Gamma(\beta+1)} + \frac{t^{2\beta}}{\Gamma(2\beta+1)} - \frac{\Gamma(2\alpha+1)t^{2\alpha+\beta}}{[\Gamma(\alpha+1)]^2\Gamma(2\alpha+\beta+1)} + \frac{t^{3\beta}}{\Gamma(3\beta+1)} + \cdots.
\end{aligned}
$$

Figure 2.4 shows the approximate solutions of y_1 and y_2 for $\alpha = 0.5,\ \beta = 0.3$.

Example 2.5 *Consider the system of nonlinear fractional differential equations [57]*

$$
\begin{cases}
D^{\alpha} y_1 = 2\, y_2^2, \\
D^{\beta} y_2 = t\, y_1, \qquad y_1(0) = 0,\ y_2(0) = 1,\ y_3(0) = 1, \\
D^{\gamma} y_3 = y_2\, y_3,
\end{cases}
$$

where $\alpha, \beta, \gamma \in (0,1)$.

In order to solve the above system, we define the nonlinear terms by

$$
N_1\bar{y} = 2\, y_2^2 = \sum_{j=0}^{\infty} A_{1j}, \quad N_2\bar{y} = x\, y_1 = \sum_{j=0}^{\infty} A_{2j}, \quad N_3\bar{y} = y_2\, y_3 = \sum_{j=0}^{\infty} A_{3j}.
$$

In view of Eq. (2.26) and using Mathematica software, we evaluate the corresponding Adomian polynomials A_{ij}, $i = 1, 2, 3$ and $j = 0, 1, \ldots$

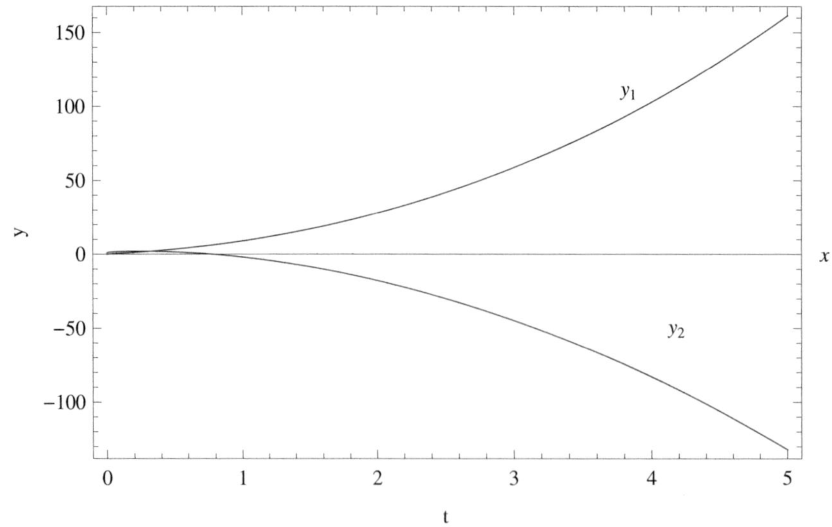

Figure 2.4 Approximate solutions of y_1 and y_2 for $\alpha = 0.5, \beta = 0.3$ using ADM for Example 2.4.

$$\begin{aligned} A_{10} &= 2\,y_{20}^2, & A_{20} &= t\,y_{10},\\ A_{11} &= 4\,y_{20}\,y_{21}, & A_{21} &= t\,y_{11},\\ A_{12} &= 2\,y_{21}^2 + 4\,y_{20}\,y_{22}, & A_{22} &= t\,y_{12},\\ A_{13} &= 4\,y_{21}y_{22} + 4\,y_{20}\,y_{23}, & A_{23} &= t\,y_{13},\\ A_{14} &= 2\,y_{22}^2 + 4\,y_{21}\,y_{23} + 4\,y_{20}\,y_{24}, & A_{24} &= t\,y_{14},\\ &\vdots & &\vdots\\ A_{30} &= y_{20}\,y_{30},\\ A_{31} &= y_{21}\,y_{30} + y_{20}\,y_{31},\\ A_{32} &= y_{22}\,y_{30} + y_{21}\,y_{31} + y_{20}\,y_{32},\\ A_{33} &= y_{23}\,y_{30} + y_{22}\,y_{31} + y_{21}\,y_{32} + y_{20}\,y_{33},\\ A_{34} &= y_{24}\,y_{30} + y_{23}\,y_{31} + y_{22}\,y_{32} + y_{21}\,y_{33} + y_{20}\,y_{34},\\ &\vdots \end{aligned}$$

The Adomian decomposition series (2.23) has the following terms:

$$\begin{cases} y_{10} = 0,\\ y_{1,m+1} = I^{\alpha} A_{1m}, \end{cases} \quad \begin{cases} y_{20} = 1,\\ y_{2,m+1} = I^{\beta} A_{2m}, \end{cases} \quad \begin{cases} y_{30} = 1,\\ y_{3,m+1} = I^{\gamma} A_{3m}, \end{cases} \quad m = 0, 1, \ldots$$

In the first iteration, we have

$$\begin{aligned} y_{11} &= I^{\alpha} A_{10} = \frac{2\,t^{\alpha}}{\Gamma(\alpha+1)},\\ y_{21} &= I^{\beta} A_{20} = 0,\\ y_{31} &= I^{\gamma} A_{30} = \frac{t^{\gamma}}{\Gamma(\gamma+1)}, \end{aligned}$$

In Figure 2.5, the approximate solutions of y_1, y_2 and y_3 are drawn for $\alpha = 0.5$, $\beta = 0.4, \gamma = 0.3$, and in Figure 2.6, the approximate solutions of y_1, y_2 and y_3 are drawn for $\alpha = \beta = \gamma = 1$, respectively.

2.2 HOMOTOPY PERTURBATION METHOD

To illustrate the basic ideas of homotopy perturbation method (HPM), we consider the following nonlinear differential equation

$$L(u) + N(u) = f(r), \quad r \in \Omega, \tag{2.28}$$

with boundary conditions

$$B(u, \partial u/\partial n) = 0, \quad r \in \Gamma, \tag{2.29}$$

where L is a linear operator, while N is nonlinear operator, B is a boundary operator, Γ is the boundary of the domain Ω and $f(r)$ is a known analytic function. The He's homotopy perturbation technique [41, 43] defines the homotopy $v(r,p) : \Omega \times [0,1] \rightarrow \Re$ which satisfies

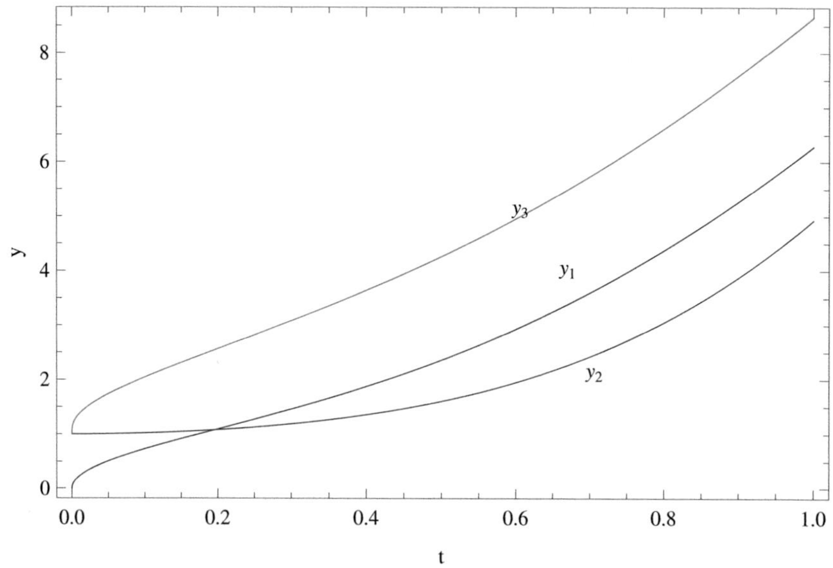

Figure 2.5 Approximate solutions of y_1, y_2 and y_3 for $\alpha = 0.5, \beta = 0.4, \gamma = 0.3$ using ADM for Example 2.5.

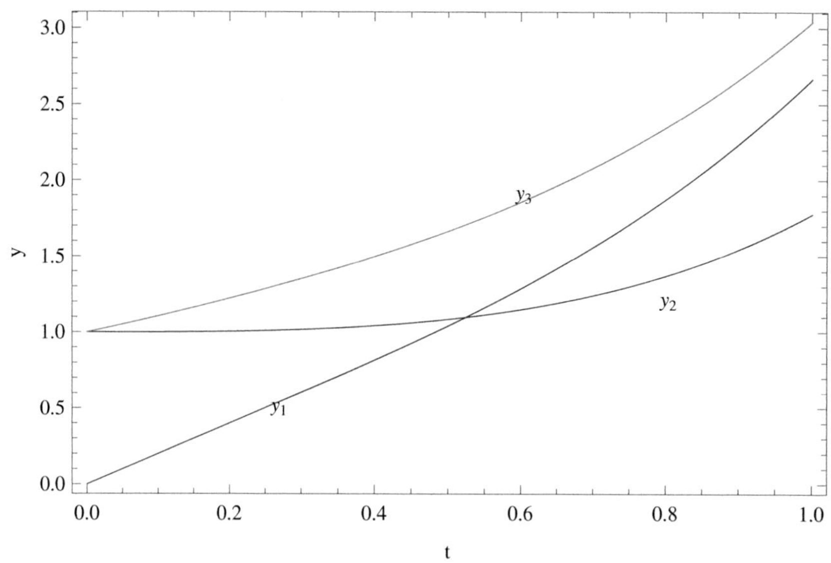

Figure 2.6 Approximate solutions of y_1, y_2 and y_3 for $\alpha = 1$, $\beta = 1$, $\gamma = 1$ using ADM for Example 2.5.

$$H(v,p) = (1-p)[L(v) - L(u_0)] + p[L(v) + N(u) - f(r)] = 0, \tag{2.30}$$

or

$$H(v,p) = L(v) - L(u_0) + pL(u_0)p[N(v) - f(r)] = 0, \tag{2.31}$$

where $r \in \Omega$ and $p \in [0,1]$ is an embedding parameter, u_0 is an initial approximation of Eq.(2.28) which satisfies the boundary conditions. Obviously, from Eqs. (2.30) and (2.31), we have

$$H(v,0) = L(v) - L(u_0) = 0,$$
$$H(v,1) = L(u_0) + N(v) - f(r) = 0.$$

The changing process of p from zero to unity is just that of $v(r,p)$ from u_0 to $u(r)$. In topology, this is called deformation, $L(v) - L(u_0)$ and $L(v) + N(v) - f(r)$ are homotopic. The basic assumption is that the solution of Eqs. (2.30) or (2.31) can be expressed as a power series in p:

$$v = v_0 + p\,v_1 + p^2\,v_2 + \cdots. \tag{2.32}$$

By substituting of in Eq. (2.32) in (2.31) and sorting the equations according to power of p we obtained a system of linear equations. By solving this system, we obtain v_i, $i = 0, 1, 2, \ldots$. It must be mentioned for nonlinear term in Eq. (2.31), we use Adomian polynomials. The approximate solution of Eq. (2.28), therefore, can be readily obtained:

$$u = \lim_{p \to 1} v = v_0 + v_1 + v_2 + \cdots. \tag{2.33}$$

Please note in some papers it is called He's polynomials! In Ref. [61], the authors have proved that these polynomials are same and obtained by the same formula.

The convergence of the series (2.33) has been proved in Refs. [41, 43].

2.2.1 Applications

Example 2.6 *Consider the following fractional Riccati equation [53] :*

$$D^\alpha u(t) = -u^2 + 1, \qquad t > 0, \quad n-1 < \alpha < n, \tag{2.34}$$

with the initial condition ,

$$u(0) = 0. \tag{2.35}$$

The exact solution, when $\alpha = 1$, is

$$u(t) = \frac{e^{2t} - 1}{e^{2t} + 1}.$$

and we can observe that, as $t \to \infty$, $u(t) \to 1$.

To solve Eq. (2.34) with initial condition (2.35) , according to the homotopy perturbation technique, we construct the following homotopy:

$$(1-p)\left(\frac{d^\alpha v}{dt^\alpha}-\frac{d^\alpha u_0}{dt^\alpha}\right)+p\left(\frac{d^\alpha v}{dt^\alpha}+v^2-1\right)=0, \tag{2.36}$$

or equivalently;

$$\frac{d^\alpha v}{dt^\alpha}-\frac{d^\alpha u_0}{dt^\alpha}+p\frac{d^\alpha u_0}{dt^\alpha}+p(v^2-1)=0, \tag{2.37}$$

Suppose the solution of Eq. (2.37) has the form

$$v=v_0+pv_1+p^2v_2+\cdots. \tag{2.38}$$

Substituting Eq. (2.38) into (2.37), and comparing coefficients of terms with identical powers of p, leads to:

$$\begin{aligned}
p^0 &: \frac{d^\alpha v_0}{dt^\alpha}-\frac{d^\alpha u_0}{dt^\alpha}=0,\\
p^1 &: \frac{d^\alpha v_1}{dt^\alpha}+\frac{d^\alpha u_0}{dt^\alpha}+v_0^2-1=0,\\
p^2 &: \frac{d^\alpha v_2}{dt^\alpha}+2v_0v_1=0,\\
p^3 &: \frac{d^\alpha v_3}{dt^\alpha}+v_1^2+2v_0v_2=0,\\
&\vdots
\end{aligned}$$

we have the initial guess $u_0(t)=\frac{t^\alpha}{\Gamma(\alpha+1)}$.

Consequently, the first few components of the homotopy perturbation for Eq. (2.37) are derived as follows:

$$\begin{aligned}
v_0(t) &= \frac{t^\alpha}{\Gamma(1+\alpha)},\\
v_1(t) &= -\frac{t^{3\alpha}\Gamma(1+2\alpha)}{\Gamma(1+\alpha)^2\Gamma(1+3\alpha)},\\
v_2(t) &= \frac{2t^{5\alpha}\Gamma(1+2\alpha)\Gamma(1+4\alpha)}{\Gamma(1+\alpha)^3\Gamma(1+3\alpha)\Gamma(1+5\alpha)},\\
v_3(t) &= -\frac{64^\alpha t^{7\alpha}\Gamma(1+2\alpha)^2\Gamma(\frac{1}{2}+3\alpha)}{\sqrt{\pi}\alpha\Gamma(\alpha)\Gamma(1+\alpha)^3\Gamma(1+3\alpha)\Gamma(1+7\alpha)}\\
&\quad-\frac{4t^{7\alpha}\Gamma(1+2\alpha)\Gamma(1+4\alpha)\Gamma(1+6\alpha)}{\alpha\Gamma(\alpha)\Gamma(1+\alpha)^3\Gamma(1+3\alpha)\Gamma(1+5\alpha)\Gamma(1+7\alpha)},\\
&\vdots
\end{aligned}$$

using the above terms, the solution $u(t)$ is:

$$\begin{aligned} u(t) &= \frac{t}{\Gamma(1+\alpha)} - \frac{t^3\Gamma(1+2\alpha)}{\Gamma(1+\alpha)^2\Gamma(1+3\alpha)} + \frac{2t^5\Gamma(1+2\alpha)\Gamma(1+4\alpha)}{\Gamma(1+\alpha)^3\Gamma(1+3\alpha)\Gamma(1+5\alpha)} \\ &- \frac{64^\alpha t^{7\alpha}\Gamma(1+2\alpha)^2\Gamma(\frac{1}{2}+3\alpha)}{\sqrt{\pi}\alpha\Gamma(\alpha)\Gamma(1+\alpha)^3\Gamma(1+3\alpha)\Gamma(1+7\alpha)} \\ &- \frac{4t^{7\alpha}\Gamma(1+2\alpha)\Gamma(1+4\alpha)\Gamma(1+6\alpha)}{\alpha\Gamma(\alpha)\Gamma(1+\alpha)^3\Gamma(1+3\alpha)\Gamma(1+5\alpha)\Gamma(1+7\alpha)} + \cdots . \end{aligned}$$

Figure 2.7 shows the solution obtained using HPM when $\alpha = 0.98$ and $\alpha = 1$ versus the exact solution . It is clear that the solution obtained from the homotopy perturbation method is consistent with the exact solution only in the neighborhood of $t = 0$. Table 2.1 shows the absolute errors of $u(t) = u_8(t)$ using HPM for the special case $\alpha = 1$.

Example 2.7 *We consider the following nonlinear system [13]:* $(0 < \alpha < 1)$

$$\begin{aligned} D_t^\alpha u + v_x w_y - v_y w_x &= -u, \\ D_t^\alpha v + w_x u_y + u_v w_x &= v, \\ D_t^\alpha w + u_x v_y + u_y v_x &= w, \end{aligned} \tag{2.39}$$

with the initial conditions as:

$$u(x,0) = e^{x+y}, \quad v(x,0) = e^{x-y}, \quad w(x,0) = e^{-x+y}, \tag{2.40}$$

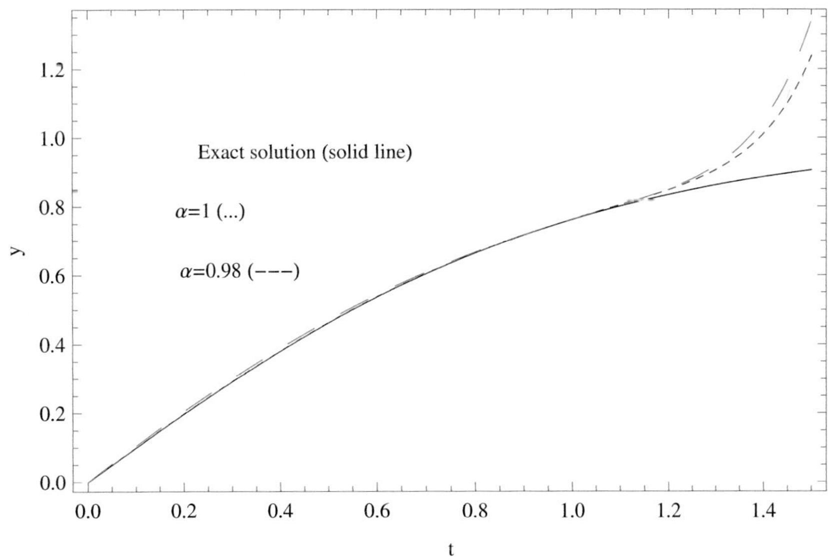

Figure 2.7 The exact solution (solid line) and homotopy perturbation method solution $u \simeq u_8$ for Example 2.6.

TABLE 2.1 Absolute Errors of $y(t) \simeq y_8(t)$ Using HPM for $\alpha = 1$ for Example 2.6

x	Absolute Errors
0	0.0000
0.1	2.42004×10^{-17}
0.2	1.87809×10^{-17}
0.3	2.6805×10^{-14}
0.4	6.17291×10^{-12}
0.5	4.14141×10^{-10}
0.6	1.27162×10^{-8}
0.7	2.27418×10^{-7}
0.8	2.73643×10^{-6}
0.9	2.43193×10^{-5}
1.0	1.70164×10^{-4}

where $0 < \alpha < 1$ *and the exact solutions, when* $\alpha = 1$ *are:*

$$u(x,t) = e^{x+y-t}, \quad v(x,t) = e^{x-y+t}, \quad w(x,t) = e^{-x+y+t}. \tag{2.41}$$

In order to solve Eq. (2.39) using HPM, we construct the following homotopy for these equations:

$$\begin{aligned} v_t &= p\left[u_t - w_x u_y - u_v w_x + v - v_t^{\alpha}\right], \\ u_t &= p\left[u_t - v_x w_y + v_y w_x - u - u_t^{\alpha}\right], \\ w_t &= p\left[w_t - u_x v_y - u_y v_x + w - wt^{\alpha}\right], \end{aligned} \tag{2.42}$$

Substituting v from Eq. (2.38) into (2.42) and rearranging based on powers of p-terms, we can obtain:

$$p^0 : \frac{\partial u_0}{\partial t} = 0, \tag{2.43}$$

$$p^0 : \frac{\partial v_0}{\partial t} = 0, \tag{2.44}$$

$$p^0 : \frac{\partial w_0}{\partial t} = 0, \tag{2.45}$$

$$\begin{aligned} p^1 : \frac{\partial u_1(x,y,t)}{\partial t} &= -\frac{\partial v_0(x,y,t)}{\partial x}\frac{\partial w_0(x,y,t)}{\partial y} + \frac{\partial v_0(x,y,t)}{\partial y}\frac{\partial w_0(x,y,t)}{\partial x} \\ &\quad - u_0(x,y,t) + \frac{\partial u_0(x,y,t)}{\partial t} - \frac{\partial^{\alpha} u_0(x,t)}{\partial t^{\alpha}}, \\ p^1 : \frac{\partial v_1(x,y,t)}{\partial t} &= -\frac{\partial w_0(x,y,t)}{\partial x}\frac{\partial u_0(x,y,t)}{\partial y} + \frac{\partial w_0(x,y,t)}{\partial y}\frac{\partial u_0(x,y,t)}{\partial x} \\ &\quad - v_0(x,y,t) + \frac{\partial v_0(x,y,t)}{\partial t} - \frac{\partial^{\alpha} v_0(x,t)}{\partial t^{\alpha}}, \end{aligned}$$

$$p^1: \frac{\partial w_1(x,y,t)}{\partial t} = -\frac{\partial v_0(x,y,t)}{\partial x}\frac{\partial u_0(x,y,t)}{\partial y} + \frac{\partial v_0(x,y,t)}{\partial y}\frac{\partial u_0(x,y,t)}{\partial x}$$
$$-w_0(x,y,t) + \frac{\partial w_0(x,y,t)}{\partial t} - \frac{\partial^\alpha w_0(x,t)}{\partial t^\alpha},$$
$$p^2: \frac{\partial u_2(x,y,t)}{\partial t} = -\frac{\partial v_0(x,y,t)}{\partial x}\frac{\partial w_0(x,y,t)}{\partial y} + \frac{\partial v_0(x,y,t)}{\partial y}\frac{\partial w_1(x,y,t)}{\partial x}$$
$$\frac{\partial v_1(x,y,t)}{\partial y}\frac{\partial w_0(x,y,t)}{\partial x} - u_1(x,y,t) + \frac{\partial u_1(x,y,t)}{\partial t} - \frac{\partial^\alpha u_1(x,t)}{\partial t^\alpha},$$

$$p^2: \frac{\partial v_2(x,y,t)}{\partial t} = -\frac{\partial u_1(x,y,t)}{\partial x}\frac{\partial w_0(x,y,t)}{\partial y} + \frac{\partial w_1(x,y,t)}{\partial y}\frac{\partial u_0(x,y,t)}{\partial x}$$
$$\frac{\partial u_1(x,y,t)}{\partial y}\frac{\partial w_0(x,y,t)}{\partial x} - \frac{\partial w_1(x,y,t)}{\partial x}\frac{\partial u_0(x,y,t)}{\partial y} - v_1(x,y,t) + \frac{\partial v_1(x,y,t)}{\partial t} - \frac{\partial^\alpha v_1(x,t)}{\partial t^\alpha},$$

$$p^2: \frac{\partial w_2(x,y,t)}{\partial t} = -\frac{\partial u_0(x,y,t)}{\partial x}\frac{\partial v_0(x,y,t)}{\partial y} + \frac{\partial v_1(x,y,t)}{\partial x}\frac{\partial u_0(x,y,t)}{\partial y}$$
$$\frac{\partial v_0(x,y,t)}{\partial y}\frac{\partial u_1(x,y,t)}{\partial x}\frac{\partial v_0(x,y,t)}{\partial x}\frac{\partial u_1(x,y,t)}{\partial y} - w_1(x,y,t) + \frac{\partial w_1(x,y,t)}{\partial t} - \frac{\partial^\alpha w_1(x,t)}{\partial t^\alpha},$$
$$\vdots$$

and therefore

$$\begin{aligned}
u_0 &= e^{x+y}, & v_0 &= e^{x-y}, & w_0 &= e^{-x+y},\\
u_1 &= -e^{x+y}\ t, & v_1 &= e^{x-y}\ t, & w_1 &= c^{-x+y}\ t,
\end{aligned}$$
$$u_2 = \frac{t^2}{2}e^{x+y}\ t^2 + \frac{e^{x+y}t^{2-\alpha}}{\Gamma 3-\alpha} - e^{x+y}\ t,$$
$$v_2 = \frac{t^2}{2}e^{x-y}\ t^2 - \frac{e^{x-y}t^{2-\alpha}}{\Gamma 3-\alpha} + e^{x-y}\ t,$$
$$w_2 = \frac{t^2}{2}e^{-x+y}\ t^2 - \frac{e^{-x+y}t^{2-\alpha}}{\Gamma 3-\alpha} + e^{-x+y}\ t,$$
$$u_3 = -\frac{t^3}{6}e^{(x+y)} + e^{(x+y)}t^2 - \frac{(t^{(3-\alpha)})^2 e^{(x+y)}}{\Gamma(4-2\alpha)t^3} - \frac{2(t-3+\alpha)t^{(3-\alpha)}e^{(x+y)}}{\Gamma(4-\alpha)t}$$
$$-e^{x+y}\ t,$$

$$
\begin{aligned}
v_3 &= \frac{t^3}{6}e^{(x-y)} + e^{(x-y)}\,t^2 + \frac{(t^{(3-\alpha)})^2\,e^{(x-y)}}{\Gamma(4-2\alpha)\,t^3} - \frac{2(t-3+\alpha)\,t^{(3-\alpha)}\,e^{(x-y)}}{\Gamma(4-\alpha)\,t} \\
&\quad - e^{x-y}\,t, \\
w_3 &= \frac{t^3}{6}e^{(-x+y)} + e^{(-x+y)}t^2 + \frac{(t^{(3-\alpha)})^2 e^{(-x+y)}}{\Gamma(4-2\alpha)t^3} - \frac{2(t-3+\alpha)t^{(3-\alpha)}e^{(-x+y)}}{\Gamma(4-\alpha)t} \\
&\quad - e^{(-x+y)}t, \\
&\vdots
\end{aligned}
$$

The solution of this equation, when $p \to 1$, will be as follows:

$$u(x,y,z,t) = u_0 + u_1 + u_2 + u_3,$$

$$v(x,y,z,t) = v_0 + v_1 + v_2 + v_3,$$

$$w(x,y,z,t) = w_0 + w_1 + w_2 + w_3,$$

Substituting $\alpha = 1$ in $u(x,y,z,t)$, $v(x,y,z,t)$ and $w(x,y,z,t)$, we ultimately obtain the solutions as below. These are exact solutions confirmed by Ref. [33].

$$u(x,y,z,t) = e^{(x+y)}\left(-t + \frac{t^2}{2!} - \frac{t^3}{3!} + \dots\right) = e^{(x+y)}e^{-t} = e^{(x+y-t)},$$

$$v(x,y,z,t) = e^{(x-y)}\left(t + \frac{t^2}{2!} + \frac{t^3}{3!} + \dots\right) = e^{(x-y)}e^{t} = e^{(x-y+t)},$$

$$w(x,y,z,t) = e^{(-x+y)}\left(t + \frac{t^2}{2!} + \frac{t^3}{3!} + \dots\right) = e^{(-x+y)}e^{t} = e^{(-x+y+t)},$$

2.2.2 Modified Homotopy Perturbation Method

Now for applying modified homotopy perturbation (MHPM), we consider the nonlinear fractional partial equation [63]

$$D_t^{\alpha} u(\bar{x},t) = Lu(\bar{x},t) + Nu(\bar{x},t) + f(x,t), \tag{2.46}$$

where ($\bar{x} = (x_1, \dots, x_n)$) and L is a linear operator, N is a nonlinear operator which also might include other fractional derivatives of order less than α, f is a known analytic function and D_t^{α}, $(n-1 < \alpha \le n, \quad n = 1,2)$ is the Caputo fractional derivative of order α subject to the initial conditions are

$$u^{(k)}(\bar{x},0) = g_k(x), \quad k = 0,\dots,n-1. \tag{2.47}$$

In view of the modified homotopy perturbation method [53], we can construct the following homotopy:

$$u^{(n)}(\bar{x},t) - Lu(\bar{x},t) - f(x,t) = p\left[u^{(n)}(\bar{x},t) + Nu(\bar{x},t) - D^{\alpha}u(\bar{x},t)\right], \quad p \in [0,1], \tag{2.48}$$

or

$$u^{(n)}(\bar{x},t) - f(x,t) = p\left[u^{(n)}(\bar{x},t) + Lu(\bar{x},t) + Nu(\bar{x},t) - D^{\alpha}u(\bar{x},t)\right], \quad p \in [0,1]. \tag{2.49}$$

The construction of the homotopy in Eqs. (2.48) or (2.49) is very similar to the technology in Ref. [40] where variational iteration method is first used to obtain approximate analytical solution for seepage flow with fractional derivatives in porous media. The homotopy parameter p always changes from zero to unity. In case $p = 0$, Eq. (2.48) becomes the linearized equation

$$u^{(n)}(\bar{x},t) = f(x,t) + Lu(\bar{x},t), \tag{2.50}$$

and Eq. (2.49) becomes the linearized equation

$$u^{(n)}(\bar{x},t) = f(x,t), \tag{2.51}$$

and when it is one, Eq. (2.48) or (2.49) turns out to be the original fractional differential equation (2.28). The basic assumption is that the solution of Eq. (2.48) or (2.49) can be written as a power series in p:

$$u = u_0 + pu_1 + p^2u_2 + p^3u_3 + \cdots. \tag{2.52}$$

Substituting Eq. (2.52) into (2.49) and equating the terms with identical powers of p, we can obtain a series of equations of the following form:

$$\begin{aligned}
p^0 &: \quad u_0^{(n)}(\bar{x},t) = f(x,t), \\
p^1 &: \quad u_1^{(n)}(\bar{x},t) = u_0^{(n)}(\bar{x},t) + Lu_0(\bar{x},t) + Hu_0(\bar{x},t) - D^{\alpha}u_0(\bar{x},t), \\
p^2 &: \quad u_2^{(n)}(\bar{x},t) = u_1^{(n)}(\bar{x},t) + Lu_1(\bar{x},t) + H(u_0,u_1) - D^{\alpha}u_1(\bar{x},t), \\
p^3 &: \quad u_3^{(n)}(\bar{x},t) = u_2^{(n)}(\bar{x},t) + Lu_2(\bar{x},t) + H(u_0,u_1,u_2) - D^{\alpha}u_2(\bar{x},t), \\
&\vdots \qquad \vdots
\end{aligned} \tag{2.53}$$

where $H(u_0, u_1, \ldots, u_j)$ depend upon $u_0, u_1, \ldots, u_j$. In view of Eq. (2.52) to determine
$N(u_0, u_1, \ldots, u_j)$, we use

$$\begin{aligned}
H(u_0,u_1,\ldots,u_j) &= \frac{1}{j!}\frac{d^j}{dp^j}\left[N\left(\sum_{i=0}^{\infty} u_j(\bar{x},t)p^j\right)\right]_{p=0} \\
&= \frac{1}{j!}\frac{d^j}{dp^j}\left[N\left(\sum_{i=0}^{j} u_j(\bar{x},t)p^j\right)\right]_{p=0}.
\end{aligned} \tag{2.54}$$

Setting $p = 1$ in Eq. (2.52) yields the solution of Eq. (2.28). It is obvious that the system of linear/nonlinear equations in Eq. (2.53) are easy to solve and the components u_j, $j \geq 0$ of the homotopy perturbation method can be completely determined and the series solutions are thus entirely determined. Finally, we approximate the solution $u(\bar{x},t) = \sum_{j=0}^{\infty} u_j(\bar{x},t)$ by the truncated series

$$u(\bar{x},t) \approx \sum_{j=0}^{S-1} u_j(\bar{x},t), \tag{2.55}$$

For the convergence of the above method, we refer the reader to the work of He [41].

2.2.2.1 *Applications*

Example 2.8 *Consider the fractional diffusion equation [63]*

$$\begin{aligned} D_t^{\alpha} u &= -\sum_{i=1}^{3} \frac{\partial^2 u}{\partial x_i^2}, \qquad -\infty < x_i < \infty, \quad t > 0, \\ u(\bar{x},0) &= e^{-(x_1+x_2+x_3)}, \qquad \alpha \in (0,1). \end{aligned} \tag{2.56}$$

In view of the modified homotopy perturbation method, the homotopy for Eq. (2.56) can be constructed as

$$u' = p\left[u' - \sum_{i=1}^{3} \frac{\partial^2 u}{\partial x_i^2} - D^{\alpha} u\right]. \tag{2.57}$$

Substituting (2.52) and the initial condition into Eq. (2.57), we obtain the following set of linear fractional differential equations:

$$\begin{aligned} u_0' &= 0, \quad u_0 = e^{-(x_1+x_2+x_3)}, \\ u_1' &= u_0' - \sum_{i=1}^{3} \frac{\partial^2 u_0}{\partial x_i^2} - D^{\alpha} u_0, \quad u_1 = 0, \\ u_2' &= u_1' - \sum_{i=1}^{3} \frac{\partial^2 u_1}{\partial x_i^2} - D^{\alpha} u_1, \quad u_2 = 0, \\ &\vdots \end{aligned} \tag{2.58}$$

Consequently, the first few components of the homotopy perturbation for Eq. (2.57) are derived as follows:

$$\begin{aligned} u_0 &= e^{-(x_1+x_2+x_3)}, \\ u_1 &= -e^{-(x_1+x_2+x_3)} \frac{t^{\alpha}}{\Gamma(\alpha+1)}, \\ &\vdots \end{aligned}$$

$$u_m = (-1)^m e^{-(x_1+x_2+x_3)} \frac{t^{m\,\alpha}}{\Gamma(m\,\alpha+1)}, \qquad m = 2, 3, \ldots,$$

using the above terms, the solution $u(\bar{x}, t)$ is:

$$\begin{aligned} u(\bar{x}, t) &= e^{-(x_1+x_2+x_3)} \left(1 - \frac{t^\alpha}{\Gamma(\alpha+1)} + \frac{t^{2\alpha}}{\Gamma(2\alpha+1)} - \frac{t^{3\alpha}}{\Gamma(3\alpha+1)} - \cdots\right) \\ &= e^{-(x_1+x_2+x_3)} \left(1 + \sum_{m=1}^{\infty} \frac{(-t^\alpha)^m}{\Gamma(m\alpha+1)}\right) = e^{-(x_1+x_2+x_3)} E_\alpha(-t), \end{aligned}$$

where E_α denotes Mittag-Leffler function.

Example 2.9 *Consider the fractional partial differential equation [63]*

$$\begin{aligned} D_t^\alpha u &= \frac{\partial^2 u}{\partial x^2} + u, \\ u(x, 0) &= \cos(\pi\,x), \qquad \alpha \in (0, 1). \end{aligned} \tag{2.59}$$

In view of the modified homotopy perturbation method (2.49), the homotopy for Eq. (2.59) can be constructed as

$$u' = p\left[u' + \frac{\partial^2 u}{\partial x^2} + u - D_t^\alpha u\right]. \tag{2.60}$$

Substituting Eq. (2.52) and the initial condition into Eq. (2.60), we obtain the following set of linear fractional differential equations:

$$\begin{aligned} u_0' &= 0, \quad u_0 = \cos(\pi\,x), \\ u_1' &= u_0' + \frac{\partial^2 u_0}{\partial x^2} + u_0 - D_t^\alpha u_0, \quad u_1 = 0, \\ u_2' &= u_1' + \frac{\partial^2 u_1}{\partial x^2} + u_1 - D_t^\alpha u_1, \quad u_2 = 0, \\ u_3' &= u_2' + \frac{\partial^2 u_2}{\partial x^2} + u_2 - D_t^\alpha u_2, \quad u_3 = 0, \\ &\vdots \end{aligned} \tag{2.61}$$

Consequently, the first few components of the homotopy perturbation for Eq. (2.59) are derived as follows:

$$\begin{aligned} u_0 &= \cos(\pi\,x), \\ u_1 &= [1-\pi^2]\cos(\pi\,x)\frac{t^\alpha}{\Gamma(\alpha+1)}, \\ u_2 &= [1-\pi^2]^2\cos(\pi\,x)\frac{t^{2\alpha}}{\Gamma(2\alpha+1)}, \\ &\vdots \end{aligned}$$

$$u_{j+1} = [1-\pi^2]^{j+1}\cos(\pi\,x)\frac{t^{j\,\alpha}}{\Gamma(j\,\alpha+1)}, \quad j=1,2,\ldots.$$

In view of the above terms,

$$\begin{aligned} u(x,t) &= \cos(\pi\,x)\left(1+\frac{[1-\pi^2]\,t^\alpha}{\Gamma(\alpha+1)}+\frac{([1-\pi^2]t^\alpha)^2}{\Gamma(2\alpha+1)}+\frac{([1-\pi^2]t^\alpha)^3}{\Gamma(3\alpha+1)}-\cdots\right)\\ &= \left(1+\sum_{m=1}^{\infty}\frac{([1-\pi^2]\,t^\alpha)^m}{\Gamma(m\alpha+1)}\right)=\cos(\pi\,x)\,E_\alpha([1-\pi^2]\,t^\alpha), \end{aligned}$$

where E_α denotes the Mittag-Leffler function.

In Figure 2.8, we draw $u(x,t)$ for $\alpha = 0.99$ and $-4 < x < 4,\ t \in [0,4]$.

Example 2.10 *Consider one-dimensional nonlinear fractional wave equation [63]:*

$$\begin{aligned} &D_t^\alpha u + a\,u_{xx} + \beta\,u + \gamma\,u^3 = 0, \qquad l_0 < x < l_1, \quad \alpha\in(1,2), \qquad (2.62)\\ &u(x,0) = B\,\tan(K\,x), \qquad u_t(x,0) = B\,c\,K\,\sec^2(K\,x), \end{aligned}$$

where a, c, β, γ, *are constant and* $B=\sqrt{\frac{\beta}{\gamma}}$, $K=\sqrt{\frac{-\beta}{2(a+c^2)}}$.

In view of Eq. (2.49), the homotopy for Eq. (2.62) can be constructed as

$$u'' = p\left[u'' + a\,u_{xx} + \beta\,u + \gamma\,u^3 + D_t^\alpha u\right]. \qquad (2.63)$$

Substituting Eq. (2.52) and the initial condition into Eq. (2.63), we obtain the following set of linear fractional differential equations:

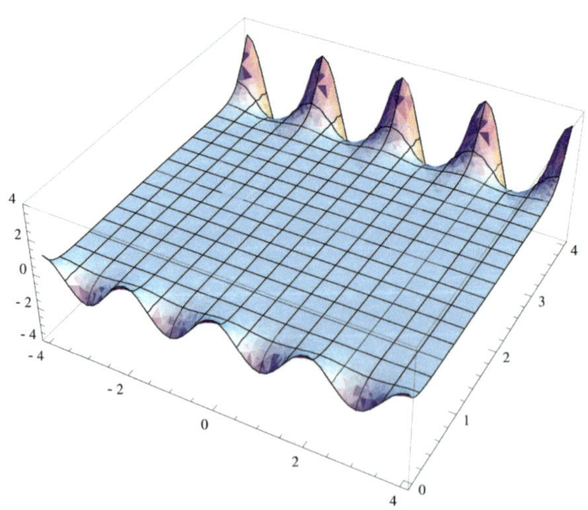

Figure 2.8 Approximate solution of $u(x,t)$ with $\alpha = 0.99$ for Example 2.9.

$$
\begin{aligned}
u_0'' &= 0, \quad u_0 = B\,\tan(K\,x)\,(u_0)_t = B\,c\,K\,\sec^2(K\,x),\\
u_1'' &= u_0'' + a\,u_{0_{xx}} + \beta\,u_0 + N(u_0) + D_t^{\alpha}u_0, \quad u_1 = 0,\\
u_2'' &= u_1'' + a\,u_{1_{xx}} + \beta\,u_1 + N(u_0,u_1) + D_t^{\alpha}u_1, \quad u_2 = 0,\\
u_3'' &= u_2'' + a\,u_{1_{xx}} + \beta\,u_2 + N(u_0,u_1,u_2) + D_t^{\alpha}u_2, \quad u_3 = 0,\\
&\vdots
\end{aligned}
\tag{2.64}
$$

where

$$
\begin{aligned}
&N(u_0) = \gamma\,u_0^3,\\
&N(u_0,u_1) = 3\gamma {u_0}^2 u_1,\\
&N(u_0,u_1,u_2) = 3\,\gamma\,u_0\,{u_1}^2 + 3\,\gamma\,{u_0}^2\,u_2.
\end{aligned}
$$

Consequently, the first few components of the homotopy perturbation for Eq. (2.62) are derived as follows:

$$
\begin{aligned}
u_0 &= B\,\tan(K\,x) + B\,c\,K\,\sec^2(K\,x)\frac{t^{\alpha}}{\Gamma(\alpha+1)},\\
u_1 &= \frac{-2\,B\,c\,K\,t^{1+\alpha}\,\beta\sec(K\,x)^2}{\Gamma(2+\alpha)} - \frac{6\,B^3\,c^3\,K^3\,t^{3+\alpha}\,\gamma\sec(K\,x)^6}{\Gamma(4+\alpha)}\\
&\quad - \frac{B\,t^{\alpha}\,\beta\,\tan(K\,x)}{\Gamma(1+\alpha)} - \frac{2\,a\,B\,K^2\,t^{\alpha}\sec(K\,x)^2\,\tan(K\,x)}{\Gamma(1+\alpha)}\\
&\quad - \frac{6\,B^3\,c^2\,K^2\,t^{2+\alpha}\,\gamma\sec(K\,x)^4\,\tan(K\,x)}{\Gamma(3+\alpha)}\\
&\quad - \frac{6\,B^3\,c\,K\,t^{1+\alpha}\,\gamma\sec(K\,x)^2\tan(K\,x)^2}{\Gamma(2+\alpha)}\\
&\quad - \frac{B^3\,t^{\alpha}\,\gamma\tan(K\,x)^3}{\Gamma(1+\alpha)}\\
&\quad - \frac{2\,aB\,c\,K\,t^{1+\alpha}\left(2\,K^2\sec(K\,x)^4 + 4\,K^2\sec(K\,x)^2\tan(K\,x)^2\right)}{\Gamma(2+\alpha)}.
\end{aligned}
$$

In Figure 2.9, $u(x,t), (\approx \Phi_4(x,t))$ is drawn for $\alpha = 1.95$, $B = 0.816497$, $c = 0.5$, $a = 2.5$, $\beta = -1.0$, $\gamma = -1.5$ and $K = 0.426401$.

Remark 2.2 *If $\alpha = 2$, the above example corresponds to the illustrative example discussed by Kaya and El-Sayed [67]. It should be remarked that the graph drawn here using modified HPM is in excellent agreement with that drawn using ADM [58].*

2.3 HOMOTOPY ANALYSIS METHOD

In 1992, Liao [75] employed the basic ideas of the homotopy in topology to propose a general analytic method for nonlinear problems, namely homotopy analysis

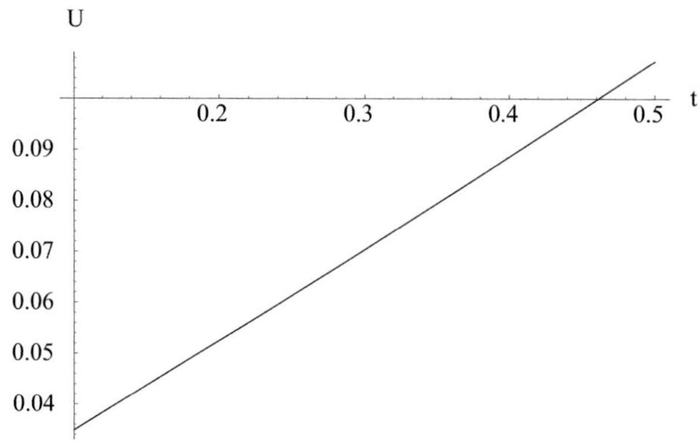

Figure 2.9 Approximate results using MHPM for Example 2.10.

method (HAM). This method has been successfully applied to solve many types of nonlinear problems [51].

We consider the following differential equation

$$\mathcal{N}[u(x,t)] = 0, \tag{2.65}$$

where $\mathcal{N}$ is a nonlinear operator, r and t are independent variables, $u(r,t)$ is an unknown function, respectively. For simplicity, we ignore all boundary or initial conditions, which can be treated in a similar way. By means of generalizing the traditional homotopy method, Liao [75] constructs the so-called zero-order deformation equation

$$(1-p)\mathcal{L}[\phi(x,t;p) - u_0(x,t)] = p\hbar H(x,t)\mathcal{N}[\phi(x,t;p)], \tag{2.66}$$

where $p \in [0,1]$ is the embedding parameter, $h \neq 0$ is a non-zero auxiliary parameter, $H(x,t) \neq 0$ is non-zero auxiliary function, $\mathcal{L} = D_t^{\alpha} \quad (n-1 < \alpha \leq n)$ is an auxiliary linear operator with the following property

$$\mathcal{L}[\phi(x,t)] = 0, \quad \text{when} \quad \phi(x,t) = 0, \tag{2.67}$$

$u_0(x,t)$ is an initial guess of $u(x,t)$, $u(x,t;p)$ is a unknown function, respectively. It is important that one has great freedom to choose auxiliary things in HAM. Obviously, when $p = 0$ and $p = 1$, it holds

$$\phi(x,t;0) = u_0(x,t), \quad \phi(x,t;1) = u(x,t), \tag{2.68}$$

respectively. Thus, as p increases from 0 to 1, the solution $\phi(x,t;p)$ varies from the initial guesses $u_0(x,t)$ to the solution $u(x,t)$. Expanding $\phi(x,t;p)$ in Taylor series with respect to p, we have

$$\phi(x,t;p) = u_0(x,t) + \sum_{m=1}^{+\infty} u_m(x,t)p^m, \tag{2.69}$$

where

$$u_m(x,t) = \frac{1}{m!}\frac{\partial^m \phi(x,t;p)}{\partial p^m}|_{p=0}. \quad (2.70)$$

If the auxiliary linear operator, the initial guess, the auxiliary parameter h, and the auxiliary function are properly chosen, the series (2.69) converges at $p = 1$, then we have

$$u_m(x,t) = u_0(x,t) + \sum_{m=1}^{+\infty} u_m(x,t), \quad (2.71)$$

Define the vector

$$\vec{u}_n = \{u_0(x,t), u_1(x,t), \ldots, u_n(x,t)\}.$$

Differentiating equation (2.66) m times with respect to the embedding parameter p and then setting $p = 0$ and finally dividing them by $m!$, we obtain the mth-order deformation equation

$$\mathcal{L}[u_m(x,t) - \chi_m u_{m-1}(x,t)] = \hbar\, H(x,t)\, R_m(\vec{u}_{m-1}, x, t), \quad (2.72)$$

where

$$R_m(\vec{u}_{m-1}, x, t) = \frac{1}{(m-1)!}\frac{\partial^{m-1}\mathcal{N}[\phi(x,t;p)]}{\partial p^{m-1}}|_{p=0}, \quad (2.73)$$

and

$$\chi_m = \begin{cases} 0, & m \leqslant 1, \\ 1, & m > 1. \end{cases} \quad (2.74)$$

Operating the Riemann-Liouville integral operator I_t^α on both side of Eq. (2.72), we have

$$u_m(x,t) = \chi_m u_{m-1}(x,t) - \chi_m \sum_{i=0}^{n-1} u_{m-1}^{(i)}(x,0^+)\frac{t^i}{i!} + \hbar H(x,t) I_t^\alpha[R_m(\vec{u}_{m-1}, x, t)]. \quad (2.75)$$

In this way, it is easily to obtain u_m for $m \geqslant 1$, at Mth order, we have

$$u(x,t) = \sum_{m=0}^{M} u_m(x,t). \quad (2.76)$$

When $M \to \infty$, we get an accurate approximation of the original equation (2.65).

2.3.1 Applications

Example 2.11 *Consider the linear fractional partial differential equation [51]:*

$$D_t^\alpha u = \frac{\partial^2 u}{\partial x^2} + u, \quad (2.77)$$

$$u(x,0) = \cos(\pi x), \qquad \alpha \in (0,1). \quad (2.78)$$

To solve Eq. (2.77) by means of homotopy analysis method, according to the initial conditions denoted in Eq. (2.78), it is natural to choose

$$u_0(x,t) = \cos(\pi x), \tag{2.79}$$

We choose the linear operator

$$\mathcal{L}[\phi(x,t;p)] = D^{\alpha}[\phi(x,t;p)], \tag{2.80}$$

with the property $\mathcal{L}[c] = 0$ where c is constant.

We now define a nonlinear operator as

$$\mathcal{N}[\phi(x,t;p)] = D_t^{\alpha}\phi(x,t;p) - \frac{\partial^2\phi(x,t;p)}{\partial x^2} - \phi(x,t;p). \tag{2.81}$$

Using the above definition, with assumption $H(x,t) = 1$, we construct the zeroth-order deformation equation

$$(1-p)\mathcal{L}[\phi(x,t;p) - u_0(x,t)] = p\hbar\mathcal{N}[\phi(r,t;p)]. \tag{2.82}$$

Obviously, when $p = 0$ and $p = 1$,

$$\phi(x,t;0) = u_0(x,t), \qquad \phi(x,t;1) = u(x,t). \tag{2.83}$$

Thus, we obtain the mth-order deformation equations

$$\mathcal{L}[u_m(x,t) - \chi_m u_{m-1}(x,t)] = \hbar R_m(\vec{u}_{m-1}), \tag{2.84}$$

where

$$R_m(\vec{u}_{m-1}) = D_t^{\alpha}u_{m-1}(x,t) - \frac{\partial^2 u_{m-1}(x,t)}{\partial x^2} - u_{m-1}(x,t).$$

Now the solution of the mth-order deformation equations (2.84)

$$u_m(x,t) = (\chi_m+\hbar)u_{m-1}(x,t) - (\chi_m+\hbar)u_{m-1}(x,0) - \hbar I_t^{\alpha}\left[\frac{\partial^2 u_{m-1}(x,t)}{\partial x^2} - u_{m-1}(x,t)\right]. \tag{2.85}$$

Finally, we have

$$u(x,t) = u_0(x,t) + \sum_{m=1}^{\infty} u_m(x,t).$$

From Eqs. (2.79) and (2.85), we obtain

$$\begin{aligned}
u_0(x,t) &= \cos(\pi x),\\
u_1(x,t) &= \frac{\hbar\pi^2 t^{\alpha}\cos(\pi x)}{\Gamma(\alpha+1)} - \frac{\hbar t^{\alpha}\cos(\pi x)}{\Gamma(\alpha+1)},\\
u_2(x,t) &= -\frac{\hbar^2\cos(\pi x)t^{\alpha}}{\Gamma(\alpha+1)} - \frac{\hbar\cos(\pi x)t^{\alpha}}{\Gamma(\alpha+1)} + \frac{\hbar^2\pi^2\cos(\pi x)t^{\alpha}}{\Gamma(\alpha+1)} + \frac{\hbar\pi^2\cos(\pi x)t^{\alpha}}{\Gamma(\alpha+1)}
\end{aligned}$$

$$+\frac{\hbar^2\cos(\pi x)t^{2\alpha}}{\Gamma(2\alpha+1)}+\frac{\hbar^2\pi^4\cos(\pi x)t^{2\alpha}}{\Gamma(2\alpha+1)}-\frac{2\hbar^2\pi^2\cos(\pi x)t^{2\alpha}}{\Gamma(2\alpha+1)},$$

$$\vdots$$

Hence

$$u(x,t)=u_0(x,t)+u_1(x,t)+u_2(x,t)+\cdots,$$

when $\hbar=-1$, we have

$$\begin{aligned}u(x,t)=&cos(\pi\ x)\left(1+\frac{(1-\pi^2)t^\alpha}{\Gamma(\alpha+1)}+\frac{((1-\pi^2)t)^{2\alpha}}{\Gamma(2\alpha+1)}\right.\\&\left.+\frac{\left(1-\pi^2\right)^3t^{3\alpha}\cos(\pi x)}{\Gamma(3\alpha+1)}+\cdots\right)\\=&cos(\pi\ x)\left(1+\sum_{m=1}^{\infty}\frac{((1-\pi^2)t^\alpha\ )^m}{\Gamma(m\alpha+1)}\right)=cos(\pi\,x)E_\alpha((1-\pi^2)t^\alpha).\end{aligned}$$

The same result was obtained using the Adomian decomposition method by Jafari and Daftardar-Gejji [58].

Example 2.12 *Consider one-dimensional nonlinear fractional wave equation [51]*

$$D_t^\alpha u+au_{xx}+\beta u+\gamma u^3=0,\ \ l_0<x<l_1,\ t>0,\qquad \alpha\in(1,2).\tag{2.86}$$

$$u(x,0)=B\tan(Kx),\quad \frac{\partial u(x,0)}{\partial t}=BcK\sec^2(Kx),\tag{2.87}$$

where a,c,β,γ *are constant and* $B=\sqrt{\frac{\beta}{\gamma}}$, $K=\sqrt{\frac{-\beta}{2(a+c^2)}}$.

To solve the Eq. (2.85) by means of homotopy analysis method, according to the initial conditions denoted in Eq. (2.87), it is natural to choose

$$u_0(x,t)=B\tan(Kx),\quad \frac{\partial u_0(x,0)}{\partial t}=BcK\sec^2(Kx).\tag{2.88}$$

We choose the linear operator

$$\mathcal{L}[\phi(x,t;p)]=D_t^\alpha[\phi(x,t;p)],\tag{2.89}$$

with the property $\mathcal{L}[C]=0$. where C is constant.

We now define a nonlinear operator as

$$\mathcal{N}[\phi(x,t;p)]=D_t^\alpha\phi(x,t;p)+a\phi(x,t;p)+\beta\phi(x,t;p)+\gamma\phi(x,t;p)^3.\tag{2.90}$$

Using above definition, with assumption $H(x,t)=1$ we construct the zeroth-order deformation equation

$$(1-p)\mathcal{L}[\phi(x,t;p)-u_0(x,t)]=p\hbar\mathcal{N}[\phi(x,t;p)].\tag{2.91}$$

Obviously, when $p = 0$ and $p = 1$,

$$\phi(x,t;0) = u_0(x,t) \quad , \quad \phi(x,t;1) = u(x,t). \tag{2.92}$$

Thus, we obtain the mth-order deformation equations

$$\mathcal{L}[u_m(x,t) - \chi_m u_{m-1}(x,t)] = \hbar R_m(\vec{u}_{m-1}), \tag{2.93}$$

where

$$R_m(\vec{u}_{m-1}) = D_t^{\alpha} u_{m-1} + a u_{m-1} + \beta u_{m-1} + \gamma \sum_{i=0}^{m-1} \left(u_i \sum_{k=0}^{m-1-i} u_k u_{m-1-i-k} \right).$$

Now the solution of the mth-order deformation equations (2.93)

$$\begin{aligned} u_m(x,t) &= (\chi_m + \hbar) u_{m-1}(x,t) - (\chi_m + \hbar) u_{m-1}(x,0) - (\chi_m + \hbar) \frac{\partial u_{m-1}(x,0)}{\partial t} t \\ &\quad + \hbar I^{\alpha} \left[a u_{m-1} + \beta u_{m-1} + \gamma \sum_{i=0}^{m-1} \left(u_i \sum_{k=0}^{m-1-i} u_k u_{m-1-i-k} \right) \right]. \end{aligned} \tag{2.94}$$

Finally, we have

$$u(x,t) = u_0(x,t) + \sum_{m=1}^{\infty} u_m(x,t).$$

From Eqs. (2.88) and (2.94) and subject to initial condition $\frac{\partial u_{m-1}(x,0)}{\partial t} = 0, m \geq 1$, we obtain

$$\begin{aligned} u_0(x,t) &= B \tan(Kx), \\ u_1(x,t) &= -bc\hbar k \sec^2(kx) t + \frac{b\hbar\beta \tan(kx) t^{\alpha}}{\Gamma(\alpha+1)} + \frac{b^3 \hbar\gamma \tan^3(kx) t^{\alpha}}{\Gamma(\alpha+1)} \\ &\quad + \frac{2ab\hbar k^2 \sec^2(kx) \tan(kx) t^{\alpha}}{\Gamma(\alpha+1)}, \\ u_2(x,t) &= -bc\hbar k \sec^2(kx) t - bc\hbar^2 k \sec^2(kx) t - \frac{4abc\hbar^2 k^3 \sec^2(kx) \tan^2(kx) t^{\alpha+1}}{(\alpha^2+\alpha)\,\Gamma(\alpha)} \\ &\quad - \frac{2abc\hbar^2 k^3 \sec^4(kx) t^{\alpha+1}}{(\alpha^2+\alpha)\,\Gamma(\alpha)} - \frac{bc\hbar^2 k\beta \sec^2(kx) t^{\alpha+1}}{(\alpha^2+\alpha)\,\Gamma(\alpha)} \\ &\quad + \frac{4^{-\alpha} b\hbar^2 \sqrt{\pi}\beta^2 \tan(kx) t^{2\alpha}}{\Gamma(\alpha+1)\Gamma\left(\alpha+\frac{1}{2}\right)} + \frac{4^{2-\alpha} a^2 b\hbar^2 k^4 \sqrt{\pi} \sec^4(kx) \tan(kx) t^{2\alpha}}{\Gamma(\alpha+1)\Gamma\left(\alpha+\frac{1}{2}\right)} \\ &\quad + \frac{4^{1-\alpha} ab\hbar^2 k^2 \sqrt{\pi}\beta \sec^2(kx) \tan(kx) t^{2\alpha}}{\Gamma(\alpha+1)\Gamma\left(\alpha+\frac{1}{2}\right)} + \frac{4^{-\alpha} b^3 \hbar^2 \sqrt{\pi}\beta\gamma \tan^3(kx) t^{2\alpha}}{\Gamma(\alpha+1)\Gamma\left(\alpha+\frac{1}{2}\right)} \\ &\quad + \frac{32^{1-2\alpha} ab^3 \hbar^2 k^2 \sqrt{\pi}\gamma \sec^4(kx) \tan(kx) t^{2\alpha}}{\Gamma(\alpha+1)\Gamma\left(\alpha+\frac{1}{2}\right)} \\ &\quad + \frac{32^{1-2\alpha} ab^3 \hbar^2 k^2 \sqrt{\pi}\gamma \sec^2(kx) \tan^3(kx) t^{2\alpha}}{\Gamma(\alpha+1)\Gamma\left(\alpha+\frac{1}{2}\right)} \end{aligned}$$

$$+\frac{2ab\hbar^2k^2\sec^2(kx)\tan(kx)t^\alpha}{\Gamma(\alpha+1)}+\frac{2^{3-2\alpha}a^2b\hbar^2k^4\sqrt{\pi}\sec^2(kx)\tan^3(kx)t^{2\alpha}}{\Gamma(\alpha+1)\Gamma\left(\alpha+\frac{1}{2}\right)}$$
$$+\frac{2ab\hbar k^2\sec^2(kx)\tan(kx)t^\alpha}{\Gamma(\alpha+1)}+\frac{b\hbar^2\beta\tan(kx)t^\alpha}{\Gamma(\alpha+1)}+\frac{b\hbar\beta\tan(kx)t^\alpha}{\Gamma(\alpha+1)}$$
$$+\frac{b^3\hbar^2\gamma\tan^3(kx)t^\alpha}{\Gamma(\alpha+1)}+\frac{b^3\hbar\gamma\tan^3(kx)t^\alpha}{\Gamma(\alpha+1)},$$

$$\vdots$$

Hence, the solution of Eq. (2.86) in series form is obtained as

$$u(x,t)=u_0(x,t)+u_1(x,t)+u_2(x,t)+\cdots,$$

Figure 2.10 shows a $\hbar$-curve, and in Figure 2.11, we draw $u(x,t)\cong u_0+u_1+u_2+u_3$ for $\alpha=1.95, \alpha=1.95,\ c=0.5,\ a=2.5,\ \beta=-1.0,\ \gamma=-1.5\ B=0.816497\ K=0.426401$ and $x=0.1$.

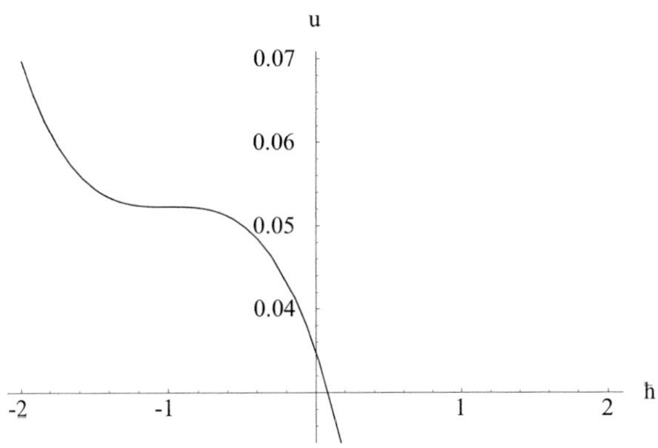

Figure 2.10 $\hbar$-curve of $u(0.1,0.1)$ for Example 2.12.

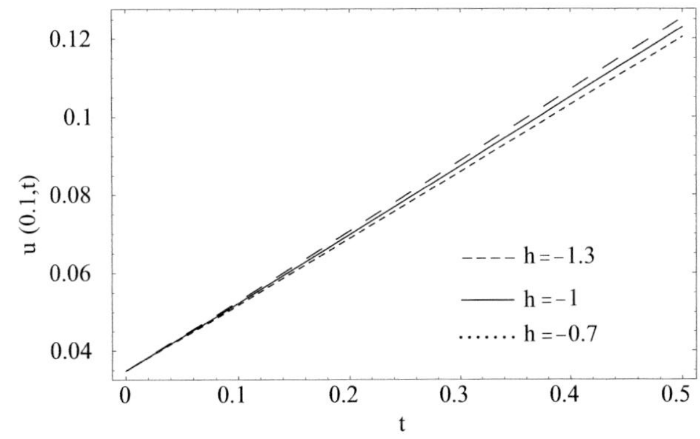

Figure 2.11 $u(0.1,t)$,using 4-term HAM solution for Example 2.12.

Remark 2.3 *This example has been solved using Adomian decomposition method [58] and modified homotopy perturbation method [63]. The graph drawn in Figure 2.11 by $\hbar = -1$ is in excellent agreement with that graph drawn with them.*

Example 2.13 *Consider the following the system of linear FPDEs:*$(0 < \alpha\,, \beta < 1)$

$$\begin{aligned} D_t^{\alpha} u - v_x + v + u &= 0\,, \\ D_t^{\beta} v - u_x + v + u &= 0, \end{aligned} \tag{2.95}$$

with initial conditions as :

$$u(x,0) = \sinh(x), \qquad v(x,0) = \cosh(x), \tag{2.96}$$

For application of homotopy analysis method, in view of the initial condition given in Eq. (2.96), it is convenient to choose

$$u_0(x,y,t) = \sinh(x), \qquad v_0(x,y,t) = \cosh(x), \tag{2.97}$$

as the initial approximate of Eq. (2.95). We choose the linear operators

$$\begin{aligned} \mathcal{L}_1[\phi_1(x,t;p)] &= D_t^{\alpha}[\phi_i(x,t;p)], \\ \mathcal{L}_2[\phi_2(x,t;p)] &= D_t^{\beta}[\phi_i(x,t;p)], \end{aligned} \tag{2.98}$$

with the property $\mathcal{L}_i[c_i] = 0$ $i = 1,2$ where c_i $(i = 1,2)$ are integral constants. We define a system of nonlinear operators as

$$\begin{aligned} \mathcal{N}_1[\phi_1,\phi_2] &= D_t^{\alpha}\phi_1 - (\phi_2)_x + (\phi_1 + \phi_2)\,, \\ \mathcal{N}_2[\phi_1,\phi_2] &= D_t^{\beta}\phi_2 - (\phi_1)_x + (\phi_1 + \phi_2)\,, \end{aligned}$$

Using the above definition, with assumption $H_i(x,t) = 1$, $i = 1,2$, we construct the zeroth-order deformation equations

$$\begin{aligned} (1-p)\mathcal{L}_1[\phi_1(x,t;p) - u_0(x,t)] &= p\hbar_1\mathcal{N}_1[\phi_1,\phi_2]\,, \\ (1-p)\mathcal{L}_2[\phi_2(x,t;p) - v_0(x,t)] &= p\hbar_2\mathcal{N}_2[\phi_1,\phi_2], \end{aligned} \tag{2.99}$$

Obviously, when $p = 0$ and $p = 1$,

$$\begin{aligned} \phi_1(x,t;0) &= u_0(x,t)\,, \quad \phi_1(x,t;1) = u(x,t), \\ \phi_2(x,t;0) &= v_0(x,t)\,, \quad \phi_2(x,t;1) = v(x,t), \end{aligned} \tag{2.100}$$

Thus, we obtain the mth-order deformation equations

$$\begin{aligned} \mathcal{L}_1[u_m(x,t) - \chi_m u_{m-1}(x,t)] &= \hbar_1 R_{1m}(\vec{u}_{m-1}, \vec{v}_{m-1}), \\ \mathcal{L}_2[v_m(x,t) - \chi_m v_{m-1}(x,t)] &= \hbar_2 R_{2m}(\vec{u}_{m-1}, \vec{v}_{m-1}), \end{aligned} \tag{2.101}$$

where

$$R_{1m}(\vec{u}_{m-1}, \vec{v}_{m-1}) = D_t^{\alpha} u_{m-1} - (v_{m-1})_x + (u_{m-1} + v_{m-1}),$$

$$R_{2m}(\vec{u}_{m-1},\vec{v}_{m-1}) = D_t^{\beta}v_{m-1} - (u_{m-1})_x + (u_{m-1}+v_{m-1}),$$

Now the solution of the mth-order deformation equations (2.101)

$$\begin{aligned} u_m(x,t) &= (\chi_m+\hbar_1)u_{m-1}(x,t) - (\chi_m+\hbar_1)u_{m-1}(x,0) \\ &\quad +\hbar_1 I^{\alpha}\left[-(v_{m-1})_x + (u_{m-1}+v_{m-1})\right], \end{aligned} \tag{2.102}$$

$$\begin{aligned} v_m(x,t) &= (\chi_m+\hbar_2)v_{m-1}(x,t) - (\chi_m+\hbar_2)v_{m-1}(x,0) \\ &\quad +\hbar_2 I^{\beta}\left[-(u_{m-1})_x + (u_{m-1}+v_{m-1})\right]. \end{aligned} \tag{2.103}$$

Finally, we have

$$u(x,t) = u_0(x,t) + \sum_{m=1}^{\infty} u_m(x,t)\,,$$

$$v(x,t) = v_0(x,t) + \sum_{m=1}^{\infty} v_m(x,t)\,.$$

From Eqs. (2.97) and (2.102) and subject to initial conditions

$$u_{m-1}(x,0)=0\,,\; v_{m-1}(x,0)=0\,,\quad m\geq 1\,,$$

we successively obtain

$$\begin{cases} u_0(x,t) = \sinh(x), \\ v_0(x,t) = \cosh(x), \end{cases} \quad \begin{cases} u_1(x,t) = \frac{t^{\alpha}\cosh(x)h_1}{\Gamma(\alpha+1)}, \\ v_1(x,t) = \frac{t^{\beta}\sinh(x)h_2}{\Gamma(\beta+1)}, \end{cases}$$

$$\begin{aligned} u_2(x,t) &= \frac{\cosh(x)\hbar_1^2 t^{\alpha}}{\Gamma(\alpha+1)} + \frac{\cosh(x)\hbar_1 t^{\alpha}}{\Gamma(\alpha+1)} + \frac{\cosh(x)\hbar_1^2 t^{2\alpha}}{\Gamma(2\alpha+1)} \\ &\quad + \frac{\sinh(x)\hbar_1 h_2 t^{2\alpha}}{\Gamma(2\alpha+1)} - \frac{\cosh(x)\hbar_1\hbar_2 t^{2\alpha}}{\Gamma(2\alpha+1)}, \\ v_2(x,t) &= \frac{\sinh(x)\hbar_2^2 t^{\beta}}{\Gamma(\beta+1)} + \frac{\sinh(x)\hbar_2 t^{\beta}}{\Gamma(\beta+1)} + \frac{\sinh(x)\hbar_2^2 t^{2\beta}}{\Gamma(2\beta+1)} \\ &\quad - \frac{\sinh(x)\hbar_1 h_2 t^{2\beta}}{\Gamma(2\beta+1)} + \frac{\cosh(x)\hbar_1\hbar_2 t^{2\beta}}{\Gamma(2\beta+1)}, \\ &\vdots \end{aligned}$$

Then the series solutions expression by HAM can be written in the form

$$\begin{aligned} u(x,t) &= u_0(x,t) + u_1(x,t) + u_2(x,t) + \cdots, \\ &= \sinh(x)\left(1 + \frac{t^{2\alpha}}{\Gamma(2\alpha+1)}(\hbar_1 h_2^2 + \hbar_1^2 h_2 + 3\hbar_1\hbar_2) + \cdots\right) \\ &\quad + \cosh(x)\left(\frac{h_1^3 t^{\alpha}}{\Gamma(\alpha+1)}(\hbar_1^3 + 3\hbar_1^2 + 3\hbar_1) + \cdots\right), \end{aligned}$$

$$
\begin{aligned}
v(x,t) &= v_0(x,t)+v_1(x,t)+v_2(x,t)+\cdots,\\
&= \sinh(x)\left(\frac{t^{\beta}}{\Gamma(\beta+1)}(\hbar_2^3+3\hbar_2^2+3\hbar_2)+\cdots\right)\\
&+\cosh(x)\left(1+\frac{t^{2\beta}}{\Gamma(2\beta+1)}(\hbar_1\hbar_2^2+\hbar_1^2\hbar_2+3\hbar_1\hbar_2)+\cdots\right),
\end{aligned}
$$

So in the special case $\hbar_i=-1$, we obtain

$$
\begin{aligned}
u(x,t) &= \sinh(x)\left(1+\frac{t^{2\alpha}}{\Gamma(2\alpha+1)}+\cdots\right)\\
&-\cosh(x)\left(\frac{t^{\alpha}}{\Gamma(\alpha+1)}+\frac{t^{3\alpha}}{\Gamma(3\alpha+1)}+\cdots\right).
\end{aligned}
$$

$$
\begin{aligned}
v(x,t) &= \cosh(x)\left(1+\frac{t^{2\beta}}{\Gamma(2\beta+1)}+\cdots\right)\\
&-\sinh(x)\left(\frac{t^{\beta}}{\Gamma(\beta+1)}+\frac{t^{3\beta}}{\Gamma(3\beta+1)}+\cdots\right).
\end{aligned}
$$

If $\alpha=\beta=1$, we get

$$
u(x,t) = \sinh\left(1+\frac{t^2}{2!}+\frac{t^4}{4!}+\cdots\right)(x)-\cosh(x)\left(t+\frac{t^3}{2!}+\frac{t^5}{5!}+\cdots\right),
$$

$$
v(x,t) = \cosh(x)\left(1+\frac{t^2}{2!}+\frac{t^4}{4!}+\cdots\right)-\sinh(x)\left(t+\frac{t^3}{3!}+\frac{t^5}{5!}+\cdots\right),
$$

which are exactly the same as the solutions obtained by variational iteration method [111] converging to the closed-form solutions, $u(x,t) = sinh(x-t), v(x,t) = cosh(x-t)$.

2.4 DAFTARDAR-GEJJI AND JAFARI METHOD

To illustrate the basic ideas of Daftardar-Gejji and Jafari Method (DJM), we consider the following general functional equation [19, 49]:

$$
y = N(y)+f, \tag{2.104}
$$

where N is a nonlinear operator from a Banach space $B \longrightarrow B$ and f a known function. We are looking for a solution y of Eq. (2.104) having the series form:

$$
y = \sum_{i=0}^{\infty} y_i. \tag{2.105}
$$

The nonlinear operator N can be decomposed as

$$
N\left(\sum_{i=0}^{\infty} y_i\right) = N(y_0)+\sum_{i=1}^{\infty}\left\{N\left(\sum_{j=0}^{i} y_j\right)-N\left(\sum_{j=0}^{i-1} y_j\right)\right\}. \tag{2.106}
$$

From Eqs. (2.105) and (2.106), Eq.(2.104) is equivalent to

$$\sum_{i=1}^{\infty} y_i = f + N(y_0) + \sum_{i=1}^{\infty} \left\{ N\left(\sum_{j=0}^{i} y_j\right) - N\left(\sum_{j=0}^{i-1} y_j\right) \right\}. \tag{2.107}$$

We define the recurrence relation:

$$\begin{cases} y_0 = f, \\ y_1 = N(y_0), \\ y_{m+1} = N(y_0 + \cdots + y_m) - N(y_0 + \cdots + y_{m-1}), \quad m = 1, 2, \ldots. \end{cases} \tag{2.108}$$

Then

$$(y_1 + \cdots + y_{m+1}) = N\left(y_0 + \cdots + y_m\right), \quad m = 1, 2, \ldots, \tag{2.109}$$

and

$$y = f + \sum_{i=1}^{\infty} y_i. \tag{2.110}$$

If N is a contraction, i.e. $||N(x) - N(y)|| \leq K||x - y||$, $0 < K < 1$, then

$$\begin{aligned} ||y_{m+1}|| &= ||N\left(y_0 + \cdots + y_m\right) - N\left(y_0 + \cdots + y_{m-1}\right)|| \\ &\leq K||y_m|| \leq K^m||y_0||, \quad m = 0, 1, 2, \ldots, \end{aligned}$$

and the series $\sum_{i=0}^{\infty} y_i$ absolutely and uniformly converges to a solution of Eq.(2.104), which is unique, in view of the Banach fixed point theorem [14]

2.4.1 Applications

Example 2.14 *Consider the following nonlinear differential equation [19]:*

$$\begin{aligned} y' &= -y^2, \\ y(1) &= 1. \end{aligned} \tag{2.111}$$

with exact solution $y = x^{-1}$ for $x > 0$.

The initial value problem in Eq. (2.111) is equivalent to the integral equation

$$y(x) = 1 - \int_1^x y^2 \, dt. \tag{2.112}$$

Following the algorithm given in

$$\begin{aligned} y_0 &= 1, \\ y_1 &= N(y_0) = -\int_1^x y_0^2 \, dt = 1 - x\,, \\ y_2 &= N(y_0 + y_1) - N(y_0) = \frac{4}{3} - 3\,x + 2\,x^2 - \frac{x^3}{3}\,, \end{aligned}$$

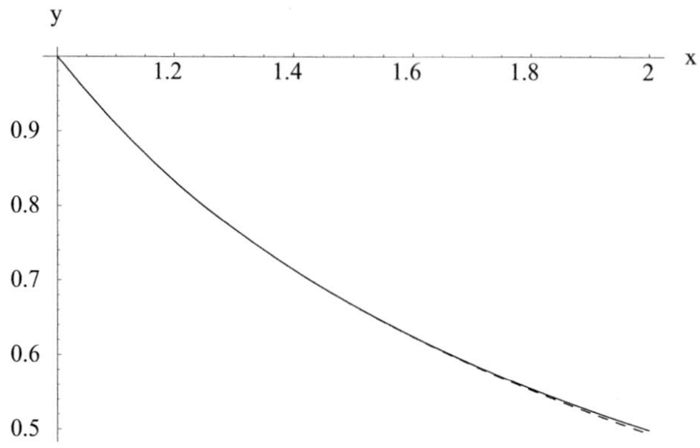

Figure 2.12 Approximate solution of $\sum_{i=0}^{5} y_i(x)$ for Example 2.14.

$$y_3 = \frac{113}{63} - \frac{64\,x}{9} + \frac{34\,x^2}{3} - \frac{85\,x^3}{9} + \frac{41\,x^4}{9} - \frac{4\,x^5}{3} + \frac{2\,x^6}{9} - \frac{x^7}{63}.$$
$$\vdots$$

We have plotted $\sum_{i=0}^{5} y_i(x)$, which is almost equal to the exact solution $y = x^{-1}$ (Figure 2.12).

Example 2.15 *Consider the following nonlinear Volterra integral equation [19]*

$$y(x) = e^x - \frac{1}{3} x\, e^{3x} + \frac{1}{3} x + \int_0^x x\, y^3(t)\, dt.$$

Here we use modified iterative scheme [110], in which we take:

$$\begin{aligned} &y_0 = e^x, \qquad y_1(x) = -\frac{1}{3} x\, e^{3x} + \frac{1}{3} x + N(y_0),\\ &y_1(x) = -\frac{1}{3} x\, e^{3x} + \frac{1}{3} x + \int_0^x x\, y_0^3(t)\, dt = 0,\\ &y_2(x) = N(y_0 + y_1) - N(y_0) = 0,\\ &y_{m+1}(x) = 0, \quad m \ge 0. \end{aligned}$$

Thus we find that the solution is $y(x) = e^x$, which is the exact solution.

Example 2.16 *Consider the system of nonlinear fractional differential equations:*

$$\begin{aligned} D^{0.5} y_1 &= 2\, y_2^2, \quad y_1(0) = 0,\\ D^{0.4} y_2 &= x\, y_1, \quad y_2(0) = 1,\\ D^{0.3} y_2 &= y_2\, y_3, \quad y_3(0) = 1. \end{aligned}$$

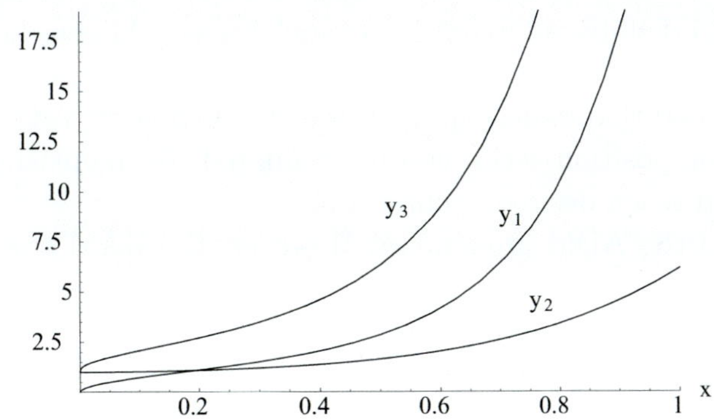

Figure 2.13 Approximate solution for DJM (five terms) for Example 2.16.

We apply a new iterative method for solving this system. The above system is equivalent to the following system of integral equations:

$$\begin{aligned} y_1 &= 2\, I^{0.5}\, y_2^2(x) = \frac{2}{\Gamma(0.5)} \int_0^x \frac{y_2^2(t)}{(x-t)^{0.5}} dt, \\ y_2 &= y_2(0) + I^{0.4}\, x\, y_1(x) = 1 + \frac{1}{\Gamma(0.4)} \int_0^x \frac{t\, y_1(t)}{(x-t)^{0.6}} dt, \\ y_3 &= y_3(0) + I^{0.3}\, (y_2(x)\, y_3(x)) = 1 + \frac{1}{\Gamma(0.3)} \int_0^x \frac{y_2(t)\, y_3(t)}{(x-t)^{0.7}} dt, \end{aligned}$$

First few terms of the iteration are given below.

$$\begin{aligned} y_{10} &= 0, \quad y_{11} = 2.25676\, x^{0.5}, \quad y_{12} = 0, \\ y_{20} &= 1, \quad y_{21} = 0, \quad y_{22} = 1.64172\, x^{1.9}, \\ y_{30} &= 1, \quad y_{31} = 1.11424\, x^{0.3}, \quad y_{32} = 1.11917\, x^{0.6}. \end{aligned}$$

This example has been solved using ADM as well [57]. In Figure 2.13, the approximate solution using the new iterative method has been plotted, respectively.

2.5 DISCUSSION AND CONCLUSION

The ADM, HPM and HAM are powerful tools that enable to find analytical/approximate solutions in case of linear and nonlinear equations. The methods have been successfully applied to both system of linear FDEs and nonlinear FDEs. These methods are powerful tools that enable one to handle even nonlinear equations. Unlike in numerical methods, these decomposition methods are free from rounding off errors and do not require large computer memory or time.

It is interesting to note that for the initial value problem $[D^{\alpha_1} y_1, \ldots, D^{\alpha_n} y_n]^t = A(y_1, \ldots, y_n)^t$, $y_i(0) = c_i, i = 1, \ldots n$, where A is a real square matrix, the solution turns out to be $\overline{y}(x) = \mathcal{E}_{(\alpha_1,\ldots,\alpha_n),1}(t^{\alpha_1} A_1, \ldots, t^{\alpha_n} A_n)\overline{y}(0)$, A_i is a matrix having ith row as $[a_{i1} \ldots a_{in}]$, and all other entries are zero and $\mathcal{E}_{(\alpha_1,\ldots,\alpha_n),1}$

$(t^{\alpha_1}A_1, \ldots, t^{\alpha_n}A_n)$ denotes an extension of multivariate Mittag-Leffler function for matrix arguments.

It is demonstrated that using capabilities of Mathematica, Adomian polynomials and terms of decomposition series can be evaluated. By increasing the number of iterations, one can reach desired accuracy.

The HPM and the ADM are similar. However, the HAM covers both of these methods.

2.5.1 Mathematica Commands

We give Mathematica commands for solving example 2.5 [49,57], using which the Adomian polynomials can be evaluated. For the sake of illustration, we consider the illustrative Example 2.5. We would like to comment here that these commands are much easier in contrast to the lengthy algorithm given in Ref. [16].

(*Define the number of iterations by specifying explicit number in place of n.*)

Itr $:= n$;

(* Define the functions Y_1, Y_2 and Y_3*)

$Y_1[\lambda\_] = \sum_{i=0}^{\text{Itr}} y_{1,i};\quad Y_2[\lambda\_] = \sum_{i=0}^{\text{Itr}} y_{2,i};\quad Y_3[\lambda\_] = \sum_{i=0}^{\text{Itr}} y_{3,i};$

(* Define the functions N_1, N_2 and N_3*)

$N_1[\lambda\_] = 2\,y_2[\lambda]\hat{}2;\ N_2[\lambda\_] = y_1[\lambda] * t;\ N_3[\lambda\_] = y_2[\lambda] * y_3[\lambda];$

$B_{1,0} = N_1[0]; A_{1,0} = B_{1,0};\ B_{2,0} = N_2[0]; A_{2,0} = B_{2,0};\ B_{3,0} = N_3[0]; A_{3,0} = B_{3,0};$

(*Calculation of Adomian polynomials*)

For $[i = 0,\ i <$ Itr, $B_{1,i} =$ Simplify $[\frac{1}{i!} *$ (Derivative $[i][N_1][\lambda])]$;

$B_{2,i} =$ Simplify $[\frac{1}{i!} *$ (Derivative $[i][N_2][\lambda])]$; $B_{3,i} =$ Simplify $[\frac{1}{i!} *$ (Derivative $[i][N_3][\lambda])]$,

$\lambda = 0; A_{1,i} = B_{1,i}/.\lambda \to 0; A_{2,i} = B_{2,i}/.\lambda \to 0; A_{3,i} = B_{3,i}/.\lambda \to 0; \lambda = .; i++]$

(* Specifying the initial conditions *)

$y_{1,0} = 0;\ y_{2,0} = 1;\ y_{3,0} = 1;\ \alpha = 0.5; \beta = 0.4; \gamma = 0.3;$

$Y_{1,0} = y_{1,0};\ Y_{2,0} = y_{2,0};\ Y_{3,0} = y_{3,0};$

(*Calculating the terms of the series*)

Do $[n = k + 1;$

$y_{1,n+1} = \frac{1}{\Gamma(\alpha)}$Integrate $[(x-t)\hat{}(\alpha-1) * A_{1,n}, (t, 0, x)$, Assumpation$\to x > 0]$;

$y_{2,n+1} = \frac{1}{\Gamma(\beta)}$Integrate$[(x-t)\hat{}(\beta-1) * A_{2,n}, (t, 0, x)$, Assumpation$\to x > 0]$;

$y_{3,n+1} = \frac{1}{\Gamma(\gamma)}$Integrate $[(x-t)\hat{}(\gamma-1) * A_{3,n}, (t, 0, x)$, Assumpation$\to x > 0]$;

$y_{1,n+1} = y_{1,n+1}/.x \longrightarrow t; y_{2,n+1} = y_{2,n+1}/.x \longrightarrow t; y_{3,n+1} = y_{3,n+1}/.x \longrightarrow t;$

$Y_{1,n+1} = Y_{1,n} + y_{1,n+1};\ Y_{2,n+1} = Y_{1,n} + y_{1,n+1};\ Y_{3,n+1} = Y_{1,n} + y_{1,n+1};$

If $[n >$ Itr, Break$[\,]], k, -1,$ Itr$]$

Print ["Y1", $n+1$, " = ", $Y_{1,n+1}$, "Y2", $n+1$, " = ", $Y_{2,n+1}$, "Y3", $n+1$, " = ", $Y_{3,n+1}$];

CHAPTER 3

Variational Iteration Method for Fractional Differential Equations

THE VARIATIONAL iteration method (VIM) was first proposed by Professor Ji-Huan He [42]. This method is an improvement on Lagrange's multiplication, and it does not require the presence of small parameters in the differential equation and provides the solution (or an approximation to it) as a sequence of iterations, which has been widely applied to solve nonlinear ordinary and partial. The variational iteration method was fully developed in 2006 and 2007 [39, 44]. Ji-Huan He [40] was the first to apply the variational iteration method to fractional differential equations. Recently, Odibat and Momani [83, 86, 88] implemented the variational iteration method to solve linear and nonlinear differential equations of fractional order.

The variational iteration method offers certain advantages over routine numerical methods. Numerical methods use discretization that gives rise to rounding off errors causing loss of accuracy and requires large computer power and time. The variational iteration method is better since it does not involve discretization of the variables, hence is free from rounding off errors and does not require large computer memory or time.

3.1 VARIATIONAL ITERATION METHOD

Here, we extend the application of the variational iteration method to solve the fractional differential equation [52, 87]:

$$D^{\alpha}y(t) + R[y(t)] + N[y(t)] = g(t), \tag{3.1}$$

where R is a linear term and N is a nonlinear one. $g(t)$ is a known function.

DOI: 10.1201/9781003474609-3

Subject to the initial and boundary conditions

$$y(0) = a, \quad 0 < \alpha \leq 1, \tag{3.2}$$

and

$$y(0) = a, \quad y'(0) = b, \quad 1 < \alpha \leq 2, \tag{3.3}$$

α $(m-1 < \alpha \leq m)$ is a parameter describing the order of the time-fractional derivative in the Caputo sense. According to the variational iteration method, we can construct the correction functional for Eq. (3.1) as:

$$\begin{aligned} y_{n+1}(t) &= y_n(t) + I^{\beta}\left[\lambda\left(D^{\alpha}y_n(t) + R[\tilde{y_n}(t)] + N[\tilde{y_n}(t)] - g(t)\right)\right] \\ &= y_n(t) + \frac{1}{\Gamma(\beta)}\int_0^t (t-\tau)^{\beta-1}\lambda(t,\tau)\left(D^{\alpha}y_n(\tau) + R[\tilde{y_n}(\tau)]\right. \\ &\quad \left. + N[\tilde{y_n}(\tau)] - g(\tau)\right) d\tau\,, \end{aligned} \tag{3.4}$$

where $\tilde{y_n}$ denotes a restricted variation. I^{β} is the Riemann-Liouville fractional integral operator of order $\beta = \alpha + 1 - \text{floor}(\alpha)$, that is $\beta = \alpha + 1 - m$ and $\lambda(t,\tau)$ is a general Lagrange multiplier, which can be identified optimally via variational theory [48]. To identify approximately Lagrange multiplier, some approximation must be made. The correction functional (3.4) can be approximately expressed as follows:

$$y_{n+1}(t) = y_n(t) + \int_0^t \lambda(t,\tau)\left(y_n^{(m)}(\tau) + R[\tilde{y_n}(\tau)] + N[\tilde{y_n}(\tau)] - g(\tau)\right) d\tau.$$

Here we apply restricted variations to y_n, in this case we can easily determine the multiplier.
Making the above functional stationary, noticing that $\delta\tilde{y_n} = 0$

$$\delta y_{n+1}(t) = \delta y_n(t) + \delta\int_{t_0}^t \lambda(t,\tau)\left(y_n^{(m)}(\tau) - g(\tau)\right) d\tau,$$

yields the following Lagrange multipliers

$$m = 1 \Longrightarrow \lambda(t,\tau) = -1, \tag{3.5}$$

$$m = 2 \Longrightarrow \lambda(t,\tau) = \tau - t. \tag{3.6}$$

Therefore, for $m = 1 (0 < \alpha \leq 1)$, we substitute $\lambda = -1$ into the functional (3.4) to obtain the following iteration formula:

$$y_{n+1}(t) = y_n(t) - I^{\alpha}\left[\left(D^{\alpha}y_n(t) + R[y_n(t)] + N[y_n(t)] - g(t)\right)\right]. \tag{3.7}$$

For $m = 2 (1 < \alpha \leq 2)$, we substitute $\tau - t$ into the functional (3.4) to get

$$\begin{aligned} y_{n+1}(t) &= y_n(t) + I^{\alpha-1}[D^\alpha y_n(t) + R[y_n(t)] + N[y_n(t)] - g(t))] \\ &= y_n(t) + \frac{1}{\Gamma(\alpha-1)} \int_0^t (t-\tau)^{\alpha-2}(\tau - t)(D_\tau^\alpha y_n(\tau) + R[y_n(\tau)] \\ &\quad + N[y_n(\tau)] - g(\tau))d\tau \\ &= y_n(t) - \frac{\alpha-1}{\Gamma(\alpha)} \int_0^t (t-\tau)^{\alpha-1}(D_\tau^\alpha y_n(\tau) + R[y_n(\tau)] \\ &\quad + N[y_n(\tau)] - g(\tau))d\tau. \end{aligned} \tag{3.8}$$

So, we obtain the following iteration formula:

$$y_{n+1}(t) = y_n(t) - (\alpha - 1)I^\alpha[D^\alpha y_n(t) + R[y_n(t)] + N[y_n(t)] - g(t)]. \tag{3.9}$$

The initial approximation (trial function) y_0 can be freely chosen if it satisfies the initial and boundary conditions of the problem. However, the success of the method depends on the proper selection of the initial approximation y_0. Finally, we approximate the solution $y(t) = \lim_{n \longrightarrow \infty} y_n(t)$ by the Nth term $y_N(t)$.

Theorem 3.1 *(Banach's Fixed Point Theorem). Assume that X is a Banach space and $A : X \to X$ is a nonlinear mapping, and suppose that*

$$\|A[u] - A[v]\| \leq \kappa \|u - v\|, \quad u, v \in X,$$

for some constants $\kappa < 1$. Then A has a unique fixed point. Furthermore, the sequence $u_{n+1} = A[u_n]$, with an arbitrary choice of $u_0 \in X$, converges to the fixed point of A.

According to Theorem 3.1, for the nonlinear mapping

$$A[y(t)] = y(t) - I^\alpha \left[D^\alpha y(t) + R[y(t)] + N[y(t)] - g(t)\right],$$

a sufficient condition for convergence of the variational iteration method is strict contraction of A. Furthermore, the sequence (3.7) converges to the fixed point of A which is also the solution of problem (3.1).

3.1.1 Applications

Example 3.1 *Consider the fractional Liénard equation [4]*

$$D^\alpha y(t) - y(t) + 4y^3(t) - 3y^5(t) = 0, \quad 1 < \alpha \leq 2, \tag{3.10}$$

with the initial conditions

$$y(0) = \frac{\sqrt{2}}{2}, \qquad y'(0) = \frac{\sqrt{2}}{4}, \tag{3.11}$$

the exact solution for $\alpha = 2$ *is*

$$y(t) = \sqrt{\frac{1+\tanh}{2}}.$$

According to the variational iteration method and to Eq. (3.9), the iteration formula for Eq. (3.10) is given by

$$\begin{aligned} y_{n+1}(t) &= y_n(t) + I^{\alpha}\left[(\tau - t)\left\{D^{\alpha}y_n(t) - y_n(t) + 4y_n^3(t) - 3y_n^5(t)\right\}\right] \\ &= y_n(t) + \frac{1}{\Gamma(\alpha-1)}\int_0^t (t-\tau)^{\alpha-2}(\tau - t)\left\{D^{\alpha}y_n(\tau) - y_n(\tau)\right. \\ &\quad \left. + 4y_n^3(\tau) - 3v_n^4(\tau)\right\} d\tau \\ &= y_n(t) - \frac{1}{\Gamma(\alpha-1)}\int_0^t (t-\tau)^{\alpha-1}\left\{D^{\alpha}y_n(\tau) - y_n(\tau)\right. \\ &\quad \left. + 4y_n^3(\tau) - 3y_n^5(\tau)\right\} d\tau. \end{aligned}$$

with $y_0(t) = \frac{\sqrt{2}}{2} + \frac{\sqrt{2}}{4}t$, Table 3.1 shows the absolute errors obtained by variational iteration method with $\alpha = 2$.

Example 3.2 *Consider the following one-dimensional linear in-homogeneous fractional wave equation [87]*

$$\frac{\partial^{\alpha} y(x,t)}{\partial t^{\alpha}} + \frac{\partial y(x,t)}{\partial x} = \frac{t^{1-\alpha}}{\Gamma(2-\alpha)}\sin x + t\cos x, \qquad 0 < \alpha \leq 1, \tag{3.12}$$

TABLE 3.1 Absolute Errors of $y_2(t)$ and $y_3(t)$ when $\alpha = 2$ for Example 3.1

t	$y_2(t)$	$y_3(t)$
0.1	3.35884×10^{-10}	6.89448×10^{-14}
0.2	1.74928×10^{-8}	1.2856×10^{-11}
0.3	1.47336×10^{-7}	2.09916×10^{-10}
0.4	5.34153×10^{-7}	1.10045×10^{-9}
0.5	1.05635×10^{-6}	2.43368×10^{-9}
0.6	1.08777×10^{-6}	1.05579×10^{-9}
0.7	7.69475×10^{-7}	6.7183×10^{-9}
0.8	6.60355×10^{-6}	2.77431×10^{-8}
0.9	4.57600×10^{-5}	1.87908×10^{-7}
1	1.95751×10^{-4}	1.34737×10^{-6}

subject to the initial condition

$$y(x,0) = 0, \tag{3.13}$$

the exact solution is

$$y(x,t) = t\sin x.$$

According to the variational iteration method and to Eq. (3.7), the iteration formula for Eq. (3.12) is given by

$$y_{n+1}(x,t) = y_n(x,t) - I_t^{\alpha}\left[\frac{\partial^{\alpha} y_n(x,t)}{\partial t^{\alpha}} + \frac{\partial y_n(x,t)}{\partial x} - \frac{t^{1-\alpha}}{\Gamma(2-\alpha)}\sin x - t\cos x\right]. \tag{3.14}$$

By the variational iteration formula (3.14), if we begin with $y_0 = 0$, we can obtain the following approximations

$$\begin{aligned}
y_1(x,t) &= y_0(x,t) - I_t^{\alpha}\left[\frac{\partial^{\alpha} y_0(x,t)}{\partial t^{\alpha}} + \frac{\partial y_0(x,t)}{\partial x} - \frac{t^{1-\alpha}}{\Gamma(2-\alpha)}\sin x - t\cos x\right] \\
&= t\sin x + \frac{t^{\alpha+1}}{\Gamma(\alpha+2)}\cos x, \\
y_2(x,t) &= y_1(x,t) - I_t^{\alpha}\left[\frac{\partial^{\alpha} y_1(x,t)}{\partial t^{\alpha}} + \frac{\partial y_1(x,t)}{\partial x} - \frac{t^{1-\alpha}}{\Gamma(2-\alpha)}\sin x - t\cos x\right] \\
&= t\sin x + \frac{t^{\alpha+1}}{\Gamma(\alpha+2)}\cos x - \frac{t^{\alpha+1}}{\Gamma(\alpha+2)}\cos x + \frac{t^{2\alpha+1}}{\Gamma(2\alpha+2)}\sin x, \\
y_3(x,t) &= y_2(x,t) - I_t^{\alpha}\left[\frac{\partial^{\alpha} y_1(x,t)}{\partial t^{\alpha}} + \frac{\partial y_2(x,t)}{\partial x} - \frac{t^{1-\alpha}}{\Gamma(2-\alpha)}\sin x - t\cos x\right] \\
&= t\sin x + \frac{t^{\alpha+1}}{\Gamma(\alpha+2)}\cos x - \frac{t^{\alpha+1}}{\Gamma(\alpha+2)}\cos x + \frac{t^{2\alpha+1}}{\Gamma(2\alpha+2)}\sin x \\
&\quad - \frac{t^{2\alpha+1}}{\Gamma(2\alpha+2)}\sin x - \frac{t^{3\alpha+1}}{\Gamma(3\alpha+2)}\cos x, \\
&\vdots
\end{aligned} \tag{3.15}$$

the exact solution is as

$$\begin{aligned}
y(x,t) &= \lim_{n\to\infty} y_n(x,t) \\
&= t\sin x + \frac{t^{\alpha+1}}{\Gamma(\alpha+2)}\cos x - \frac{t^{\alpha+1}}{\Gamma(\alpha+2)}\cos x + \frac{t^{2\alpha+1}}{\Gamma(2\alpha+2)}\sin x \\
&\quad - \frac{t^{2\alpha+1}}{\Gamma(2\alpha+2)}\sin x - \frac{t^{3\alpha+1}}{\Gamma(3\alpha+2)}\cos x + \cdots.
\end{aligned}$$

By canceling the noise terms and keeping the non-noise terms yield the exact solution of Eq. (3.12). If we begin with $y_0 = t\sin(x)$, then the exact solution follows immediately by using two iterations.

$$\begin{aligned} y_1(x,t) &= y_0(x,t) - I_t^{\alpha}\left[\frac{\partial^{\alpha} y_0(x,t)}{\partial t^{\alpha}} + \frac{\partial y_0(x,t)}{\partial x} - \frac{t^{1-\alpha}}{\Gamma(2-\alpha)}\sin x - t\cos x\right] \\ &= t\sin x - I_t^{\alpha}\left[\frac{t^{1-\alpha}}{\Gamma(2-\alpha)}\sin x + t\cos x - \frac{t^{1-\alpha}}{\Gamma(2-\alpha)}\sin x - t\cos x\right] \\ &= t\sin x. \end{aligned} \tag{3.16}$$

Example 3.3 *We consider the time-fractional KdV equation [87]*

$$\frac{\partial^{\alpha} y}{\partial t^{\alpha}} + 6y\frac{\partial y}{\partial x} + \frac{\partial^3 y}{\partial x^3} = 0, \qquad 0 < \alpha \le 1, \tag{3.17}$$

with the initial condition

$$y(x,0) = \frac{1}{2}\operatorname{sech}^2\left(\frac{1}{2}x\right), \tag{3.18}$$

the exact solution, for the special case $\alpha = 1$, is given by

$$y(x,t) = \frac{1}{2}\operatorname{sech}^2\left(\frac{1}{2}(x-t)\right).$$

According to the variational iteration method and to Eq. (3.7), the iteration formula for the time-fractional KdV Eq. (3.17) is given by

$$y_{n+1}(x,t) = y_n(x,t) - I_t^{\alpha}\left[\frac{\partial^{\alpha} y_n(x,t)}{\partial t^{\alpha}} + 6y_n\frac{\partial y_n}{\partial x} + \frac{\partial^3 y_n}{\partial x^3}\right]. \tag{3.19}$$

By the variational iteration formula (3.19), if we begin with $y_0(x,t) = \frac{1}{2}\operatorname{sech}^2(\frac{1}{2}x)$, we can obtain the following approximations

$$\begin{aligned} y_1(x,t) &= y_0(x,t) - I_t^{\alpha}\left[\frac{\partial^{\alpha} y_0}{\partial t^{\alpha}} + 6y_0\frac{\partial y_0}{\partial x} + \frac{\partial^3 y_0}{\partial x^3}\right] \\ &= \frac{1}{2}\operatorname{sech}^2\left(\frac{1}{2}x\right) - \left[\frac{3}{2}\operatorname{sech}^4\left(\frac{1}{2}x\right)\tanh\left(\frac{1}{2}x\right)\right. \\ &\quad \left. + \operatorname{sech}^4\left(\frac{1}{2}x\right)\tan h\left(\frac{1}{2}x\right) - \frac{1}{2}\operatorname{sech}^2\left(\frac{1}{2}x\right)\tan h^3\left(\frac{1}{2}x\right)\right]\frac{t^{\alpha}}{\Gamma(\alpha+1)}. \end{aligned}$$

The absolute errors of the fourth-order term approximate solution for Eq. (3.17) obtained using the variational iteration method for $\alpha = 1$ and $t = 0.1$ are given in Table 3.2. Figure 3.1 show exact solution and the solution obtained of Eq. (3.17) for $t = 0.1$ and the different values of α.

TABLE 3.2 Absolute Errors of $y(x,t)$ for $t = 0.1$ and $\alpha = 1$ for Example 3.3

(x,t)	$\alpha = 1$
0.1	6.11371×10^{-6}
0.2	9.43692×10^{-6}
0.3	1.15168×10^{-5}
0.4	1.19775×10^{-5}
0.5	1.06436×10^{-5}
0.6	7.055181×10^{-6}
0.7	2.93308×10^{-6}
0.8	2.83096×10^{-6}
0.9	9.26612×10^{-6}
1	1.58687×10^{-5}

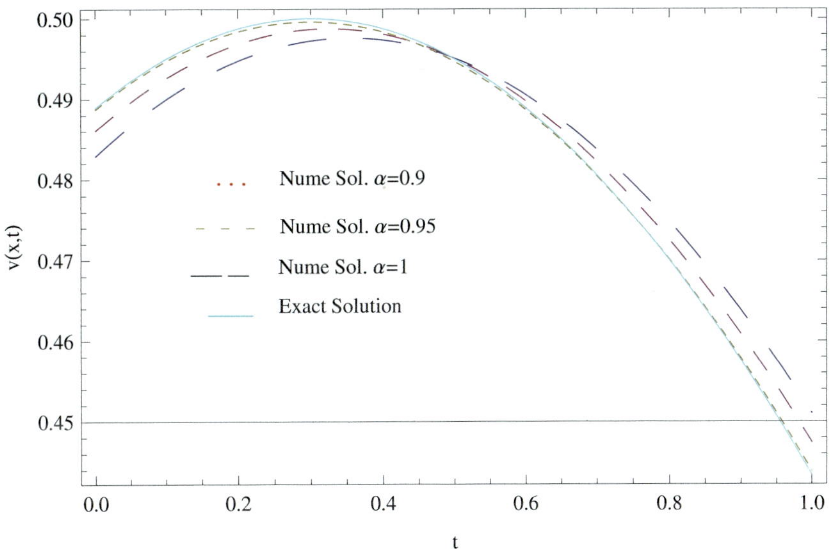

Figure 3.1 Comparison of the exact solution and the approximate solutions in $t = 0.3$ for different values of α for Example 3.3.

3.2 MODIFIED VARIATIONAL ITERATION METHOD

The main drawback of the standard VIM is that the sequence of successive approximations of the solution obtained can be rapidly convergent only in a small region, which will greatly restrict the application area of such a method. To enlarge the convergence region of the sequence of successive approximations obtained, we shall modify the VIM by introducing an auxiliary parameter [64] For Eq. (3.1), we rewrite it as

$$D^{\alpha}y(t) - D^{\alpha}y(t) + \gamma\left[D^{\alpha}y(t) + R(y(t)) + N(y(t)) - g(t)\right] = 0, \tag{3.20}$$

where γ is an auxiliary parameter and $\gamma \neq 0$, which is used to adjust the convergence region of the following iterative formula.

A correct functional for Eq. (3.20) can be written as

$$\begin{aligned} y_{n+1}(t) &= y_n(t) \\ &+ I^{\alpha}\lambda\left\{D^{\alpha}y_n(t) - D^{\alpha}\tilde{y_n}(t) + \gamma\left[D^{\alpha}\tilde{y_n}(t) + R(\tilde{y_n}(t)) + N(\tilde{y_n}(t)) - g(t)\right]\right\}, \end{aligned} \tag{3.21}$$

where $\tilde{y_n}$ is a restricted variation, i.e. $\delta\tilde{y_n} = 0$; λ is a general Lagrangian multiplier and according to the VIM, the following iteration formula for modified variational iteration method (MVIM) can be obtained

$$y_{n+1}(t) = y_n(t) + \gamma I^{\alpha}\lambda\left[D^{\alpha}y_n(t) + R(y_n(t)) + N(y_n(t)) - g(t)\right]. \tag{3.22}$$

From the convergence analysis in Section 3.1, it is easy to see that the smaller the value of $|\gamma|$ is, the wider the convergence region of iterative sequence (3.22) is. In fact, iterative formula (3.22) gives us vast freedom of choice. For some strong nonlinear problems, one can choose a relatively small value of $|\gamma|$ (generally less than 1) to obtain a good approximation in a wider region. In addition, it should be especially pointed out that when $|\gamma| = 1$, (3.22) becomes the standard variational iteration formula (3.7).

3.2.1 Applications

Example 3.4 *Consider the following fractional Riccati equation [52, 64]*

$$D^{\alpha}y(t) = 1 + 2y(t) - y^2(t), \qquad 0 < \alpha < 1, \quad 0 \le t \le 4, \tag{3.23}$$

with the initial condition

$$y(0) = 0, \tag{3.24}$$

the exact solution when $\alpha = 1$ is

$$y(t) = 1 + \sqrt{2}\tan h\left[\sqrt{2}t + \frac{1}{2}\log\left(\frac{-1+\sqrt{2}}{1+\sqrt{2}}\right)\right]. \tag{3.25}$$

The VIM for Eq. (3.23) leads to the following iteration formula:

$$y_{n+1}(t) = y_n(t) - I^{\alpha}\left[D^{\alpha}y_n(t) - 1 - 2y_n(t) + y_n^2(t)\right]. \tag{3.26}$$

Beginning with an initial approximation $y_0(t) = \frac{t^{\alpha}}{\Gamma(\alpha+1)}$, we obtain the following successive approximations

$$\begin{aligned} y_1(t) &= \frac{t^{\alpha}}{\Gamma(\alpha+1)} + \frac{2t^{2\alpha}}{\Gamma(2\alpha+1)} - \frac{\Gamma(2\alpha+1)t^{3\alpha}}{\Gamma^2(\alpha+1)\Gamma(3\alpha+1)}, \\ y_2(t) &= \frac{t^{\alpha}}{\Gamma(\alpha+1)} + 2\frac{t^{2\alpha}}{\Gamma(2\alpha+1)} + \left\{4 + \frac{\Gamma(2\alpha+1)}{\Gamma^2(\alpha+1)}\right\}\frac{t^{3\alpha}}{\Gamma(3\alpha+1)} \end{aligned}$$

$$+ \quad \frac{4\Gamma(3\alpha+1)}{\Gamma(\alpha+1)\Gamma(2\alpha+1)} - \frac{2\Gamma(2\alpha+1)}{\Gamma^2(\alpha+1)} \times \frac{t^{4\alpha}}{\Gamma(4\alpha+1)}$$
$$+ \quad \left(\frac{4}{\Gamma^2(2\alpha+1)} - \frac{2\Gamma(2\alpha+1)}{\Gamma^3(\alpha+1)\Gamma(3\alpha+1)}\right) \times \frac{\Gamma(t^\alpha+1)t^{5\alpha}}{\Gamma(5\alpha+1)}$$
$$- \quad 4\frac{\Gamma(5\alpha+1)t^{6\alpha}}{\Gamma^2(\alpha+1)\Gamma(3\alpha+1)\Gamma(6\alpha+1)} + \frac{\Gamma^2(2\alpha+1)\Gamma(6\alpha+1)t^{7\alpha}}{\Gamma^4(\alpha+1)\Gamma^2(3\alpha+1)\Gamma(7\alpha+1)}$$
$$\vdots$$

Figure 3.2 shows comparison between the exact solution and the approximate solutions of $y_5(t)$ obtained using VIM for the different values of α.

According to the MVIM, we consider the correction functional in the following form

$$\begin{aligned} y_{n+1}(t) &= y_n(t) + \frac{1}{\Gamma(\alpha)} \int_0^t (t-\tau)^{\alpha-1} \lambda(\tau) \Big\{ D^\alpha y_n(\tau) - D^\alpha \tilde{y_n}(\tau) \\ &+ \gamma \Big[D^\alpha \tilde{y_n}(\tau) - 1 - 2\tilde{y_n}(\tau) + \tilde{y_n}^2(\tau) \Big] \Big\} d\tau, \end{aligned} \tag{3.27}$$

The function $\tilde{y_n}$ is a restricted variation, which means $\delta \tilde{y_n} = 0$. According to the mentioned method, the Lagrange multiplier, can easily be identified as

$$\lambda = -1.$$

So, we have the following iteration formula

$$y_{n+1}(t) = y_n(t) - I^\alpha \gamma \{ D^\alpha y_n(t) - 1 - 2y_n(t) + y_n^2(t) \}. \tag{3.28}$$

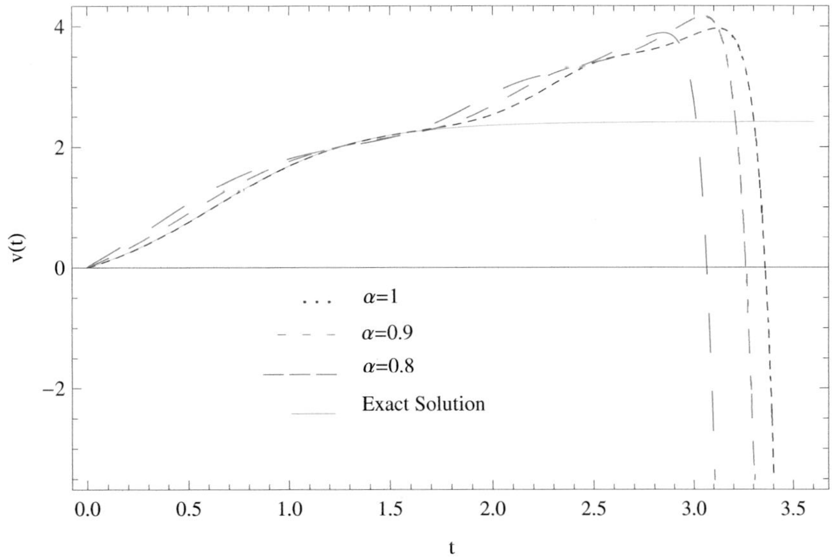

Figure 3.2 Comparison of the exact solution and the approximate solutions $y_5(t)$ using VIM for the different values of α for Example 3.4.

Beginning with an initial approximation $y_0(t) = \frac{t^\alpha}{\Gamma(\alpha+1)}$, we obtain the following successive approximations

$$
\begin{aligned}
y_1(t) &= \frac{t^\alpha}{\Gamma(\alpha+1)} + \frac{2\gamma t^{2\alpha}}{\Gamma(2\alpha+1)} - \frac{\gamma\Gamma(2\alpha+1)t^{3\alpha}}{\Gamma^2(\alpha+1)\Gamma(3\alpha+1)}, \\
y_2(t) &= \frac{t^\alpha}{\Gamma(\alpha+1)} + \frac{2\gamma t^{2\alpha}}{\Gamma(2\alpha+1)} + \left(4 + \frac{\gamma\Gamma(2\alpha+1)}{\Gamma^2(\alpha+1)}\right)\frac{t^{3\alpha}\gamma^2}{\Gamma(3\alpha+1)} \\
&+ \left(\frac{4\Gamma(3\alpha+1)}{\Gamma(\alpha+1)\Gamma(2\alpha+1)} - \frac{2\Gamma(2\alpha+1)}{\Gamma^2(\alpha+1)}\right) \times \frac{t^{4\alpha}\gamma^2}{\Gamma(4\alpha+1)} \\
&+ \left(\frac{4}{\Gamma^2(2\alpha+1)} - \frac{2\Gamma(2\alpha+1)\gamma^2}{\Gamma^3(\alpha+1)\Gamma(3\alpha+1)}\right) \times \frac{\Gamma(t^\alpha+1)t^{5\alpha}\gamma^2}{\Gamma(5\alpha+1)} \\
&- \frac{\Gamma(5\alpha+1)t^{6\alpha}\gamma^3}{\Gamma^2(\alpha+1)\Gamma(3\alpha+1)\Gamma(6\alpha+1)} + \frac{\Gamma^2(2\alpha+1)\Gamma(6\alpha+1)t^{7\alpha}\gamma^3}{\Gamma^4(\alpha+1)(3\alpha+1)\Gamma(7\alpha+1)}.
\end{aligned} \tag{3.29}
$$

Taking $\gamma = 0.59$ and $n = 5$, we can obtain the approximation solutions of Eq. (3.23) using MVIM on $[0, 4]$. The numerical results are shown in Figure 3.3. We find that the solution derived by the VIM gives a good approximation only in the neighborhood of the initial position, while the MVIM gives a good approximation in a wider region.

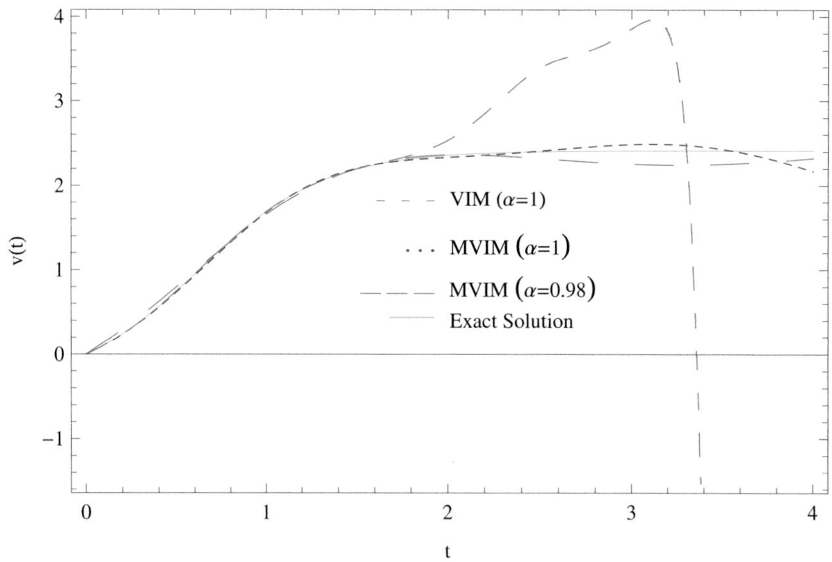

Figure 3.3 Comparison of the exact solution and the approximate solutions $y_5(t)$ using VIM and MVIM for Example 3.4.

3.3 REVISED VARIATIONAL ITERATION METHOD

In this section, a modification of the variational iteration method for solving systems of nonlinear fractional-order differential equations is proposed.

Let us consider the following system of fractional differential equations [118]:

$$\begin{cases} D^{\alpha_1} y_1(t) + N_1(y_1(t), \dots, y_m(t)) = g_1(t), \\ D^{\alpha_2} y_2(t) + N_2(y_1(t), \dots, y_m(t)) = g_2(t), \\ \vdots \\ D^{\alpha_m} y_m(t) + N_m(y_1(t), \dots, y_m(t)) = g_m(t), \end{cases} \tag{3.30}$$

with initial conditions

$$y_i(0) = d_i, \qquad i = 1, ..., m, \tag{3.31}$$

where $0 < \alpha_i \leq 1$, $i = 1, \dots, m$, N_i are operators with respect to and $g_i(t)$ are known functions. In this case, the iteration formula is obtained as follows:

$$\begin{cases} y_{1(n+1)}(t) = y_{1n}(t) - I^{\alpha_1}[D^{\alpha_1} y_{1n}(t) + N_1(y_{1n}(t), \dots, y_{mn}(t)) - g_1(t)], \\ y_{2(n+1)}(t) = y_{2n}(t) - I^{\alpha_2}[D^{\alpha_2} y_{2n}(t) + N_2(y_{1n}(t), \dots, y_{mn}(t)) - g_2(t)], \\ \vdots \\ y_{m(n+1)}(t) = y_{mn}(t) - I^{\alpha_m}[D^{\alpha_m} y_2(t) + N_m(y_{mn}(t), \dots, y_{mn}(t)) - g_m(t)]. \end{cases} \tag{3.32}$$

Here we construct the following iteration formula instead of the iteration formula obtained with the standard variational iteration method equation (3.32):

$$\begin{cases} y_{1(n+1)}(t) = y_{1n}(t) - I^{\alpha_1}[D^{\alpha_1} y_{1n}(t) + N_1(y_{1n}(t), \dots, y_{mn}(t)) - g_1(t)], \\ y_{2(n+1)}(t) = y_{2n}(t) - I^{\alpha_2}[D^{\alpha_2} y_{2n}(t) + N_2(y_{1n+1}(t), y_{1n}(t), \dots, y_{mn}(t)) - g_2(t)], \\ \vdots \\ y_{m(n+1)}(t) = y_{mn}(t) - I^{\alpha_m}[D^{\alpha_m} y_{mn}(t) + N_m(y_{1n+1}(t), \dots, y_{(m-1)(n+1)}(t), y_{mn}(t)) \\ - g_m(t)]. \end{cases} \tag{3.33}$$

In fact, the updated values $y_{1(n+1)}(t), y_{2(n+1)}(t), \dots, y_{(i-1)(n+1)}$ are used for finding $y_{i(n+1)}(t)$. We called it the revised variational iteration method (RVIM). This technique can accelerate the convergence of iterative approximate solutions relative to the approximate solutions obtained using the traditional variational iteration method. The effect of this correction is clear in $y_{i(n+1)}(t)$ because the updated values are used to compute it. Tatari and Dehghan have employed this technique for systems of ordinary differential equations [106].

3.3.1 Applications

Example 3.5 *Consider the following system of nonlinear fractional-order Van der Pol [118]*

$$\begin{cases} D^{\alpha}u(t) = v(t), & 0 < \alpha \in 1, \\ \frac{dv(t)}{dt} = -u(t) - 0.5u^2(t)v(t) + 0.5v(t), \end{cases} \tag{3.34}$$

with the initial condition

$$u(0) = 0, \qquad v(0) = 1. \tag{3.35}$$

The standard VIM for Eq. (3.34) leads to the following iteration formula:

$$\begin{cases} u_{n+1}(t) = u_n(t) - I^{\alpha}[D^{\alpha}u_n(t) - v_n(t)], \\ v_{n+1}(t) = v_n(t) - \int_0^t \left[\frac{dv_n(\tau)}{d\tau} + u_n(\tau) + 0.5u_n^2(\tau)v_n(\tau) - 0.5v_n(\tau)\right] d\tau. \end{cases} \tag{3.36}$$

Using the RVIM, we obtain

$$\begin{cases} u_{n+1}(t) = u_n(t) - I^{\alpha}[D^{\alpha}u_n(t) - v_n(t)], \\ v_{n+1}(t) = v_n(t) - \int_0^t \left[\frac{dv_n(\tau)}{d\tau} + u_{n+1}(\tau) + 0.5u_{n+1}^2(\tau)v_n(\tau) - 0.5v_n(\tau)\right] d\tau. \end{cases} \tag{3.37}$$

Starting with the initial approximations $u(0) = 0$ and $v(0) = 1$, we can easily obtain the results using Eq. (3.36) .

$$\begin{aligned} u_1(t) &= \frac{t^{\alpha}}{\Gamma(\alpha+1)}, \\ v_1(t) &= 1 + 0.5t, \\ u_2(t) &= \frac{t^{\alpha}}{\Gamma(\alpha+1)} + \frac{0.5t^{\alpha+1}}{\Gamma(\alpha+2)}, \\ v_2(t) &= 1 + 0.5t + 0.12t^2 - \frac{t^{\alpha+1}}{\Gamma(\alpha+2)} - \frac{0.5t^{2\alpha+1}}{(2\alpha+1)\Gamma^2(\alpha+1)} - \frac{0.25t^{2\alpha+2}}{(2\alpha+2)\Gamma^2(\alpha+1)}, \\ &\vdots \end{aligned}$$

Also, results using Eq. (3.37) can be obtained as

$$\begin{aligned} u_1(t) &= \frac{t^{\alpha}}{\Gamma(\alpha+1)}, \\ v_1(t) &= 1 + 0.5t - \frac{t^{\alpha+1}}{\Gamma(\alpha+2)} - \frac{0.5t^{2\alpha+1}}{(2\alpha+1)\Gamma^2(\alpha+1)}, \\ u_2(t) &= \frac{t^{\alpha}}{\Gamma(\alpha+1)} + \frac{0.5t^{\alpha+1}}{\Gamma(\alpha+2)} - \frac{t^{2\alpha+1}}{\Gamma(2\alpha+2)} - \frac{0.5t^{3\alpha+1}\Gamma(2\alpha+2)}{(2\alpha+1)\Gamma^2(\alpha+1)\Gamma(3\alpha+3)}, \\ &\vdots \end{aligned} \tag{3.38}$$

TABLE 3.3 Absolute Errors of $u(t) = u_5(t)$ Using VIM and RVIM for the Special Case $\alpha = 1$ for Example 3.5

t	$(u_5(t))$ **VIM**	$(u_5(t))$ **RVIM**
0.2	1.21288×10^{-7}	6.27721×10^{-9}
0.4	8.95826×10^{-6}	8.88751×10^{-9}
0.6	1.17174×10^{-4}	2.76336×10^{-8}
0.8	7.02546×10^{-4}	1.19925×10^{-8}
1	2.57257×10^{-3}	2.00543×10^{-7}
1.2	6.43624×10^{-3}	2.82358×10^{-7}
1.4	1.14028×10^{-2}	2.92631×10^{-7}
1.6	1.39911×10^{-2}	1.788×10^{-7}
1.8	1.04306×10^{-2}	9.35721×10^{-8}
2	8.60948×10^{-4}	3.69521×10^{-7}

TABLE 3.4 Absolute Errors of $v(t) = v_4(t)$ Using VIM and RVIM for the Special Case $\alpha = 1$ for Example 3.5

t	$v_4(t)$ **VIM**	$(v_4(t))$ **RVIM**
0.2	3.56555×10^{-9}	3.92987×10^{-9}
0.4	1.41285×10^{-9}	6.7004×10^{-8}
0.6	1.23525×10^{-3}	2.91998×10^{-7}
0.8	5.32791×10^{-3}	6.65026×10^{-7}
1	1.41058×10^{-2}	9.4798×10^{-7}
1.2	2.39152×10^{-2}	6.37843×10^{-7}
1.4	2.2661×10^{-2}	4.9675×10^{-7}
1.6	4.27141×10^{-4}	1.85184×10^{-6}
1.8	3.47783×10^{-2}	2.70134×10^{-6}
2	6.03232×10^{-2}	1.94395×10^{-6}

Tables 3.3 and 3.4 show the absolute errors of $u(t) = u_5(t)$ and $v(t) = v_4(t)$ using VIM and RVIM for the special case $\alpha = 1$, respectively. Figures 3.4 and 3.5 show comparison between the approximate solutions $u(t) = u_5(t)$ and $v(t) = v_5(t)$ of Eq. (3.30) obtained using VIM and RVIM for the special case $\alpha = 1$ and the numerical solutions for the special case $\alpha = 1$, respectively. Figures 3.6 and 3.7 show approximate solutions $u(t) = u_5(t)$ and $v(t) = v_4(t)$ of Eq. (3.30) using VIM and RVIM for the special case $\alpha = 0.98$ and the numerical solutions, respectively.

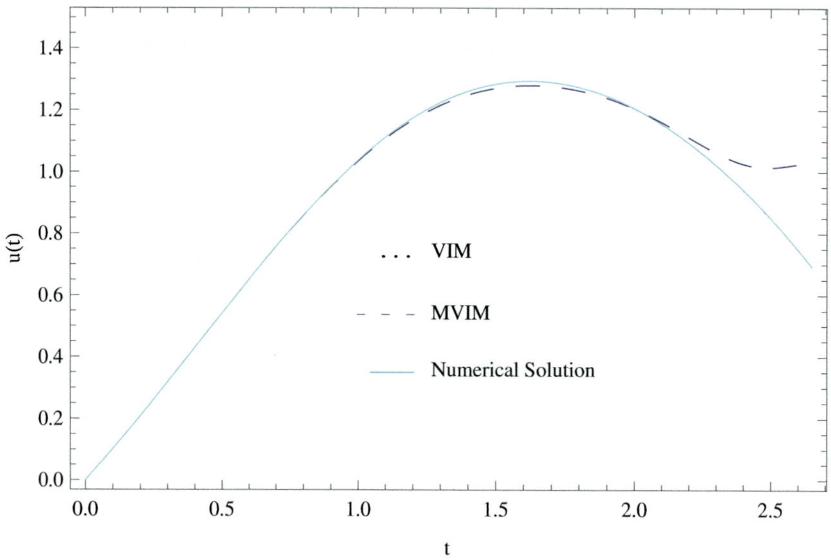

Figure 3.4 Simulation of the approximate results for $u(t) = u_5(t)$ and $\alpha = 1$ for Example 3.5.

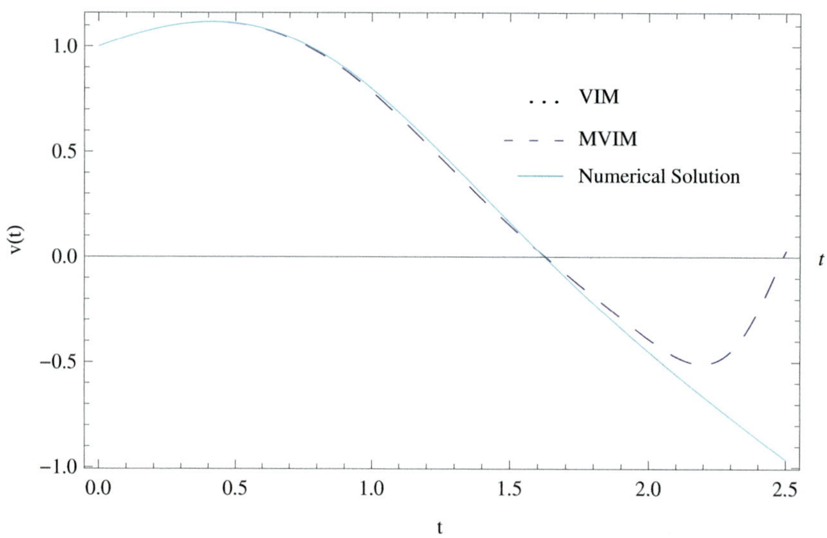

Figure 3.5 Simulation of the approximate results for $v(t) = v_4(t)$ and $\alpha = 1$ for Example 3.5.

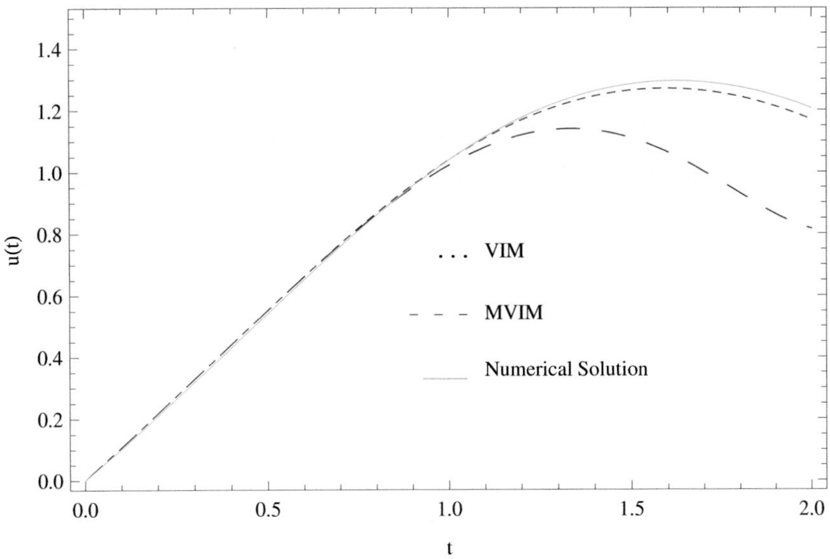

Figure 3.6 Simulation of the approximate results for $u(t) = u_5(t)$ and $\alpha = 0.98$ for Example 3.5.

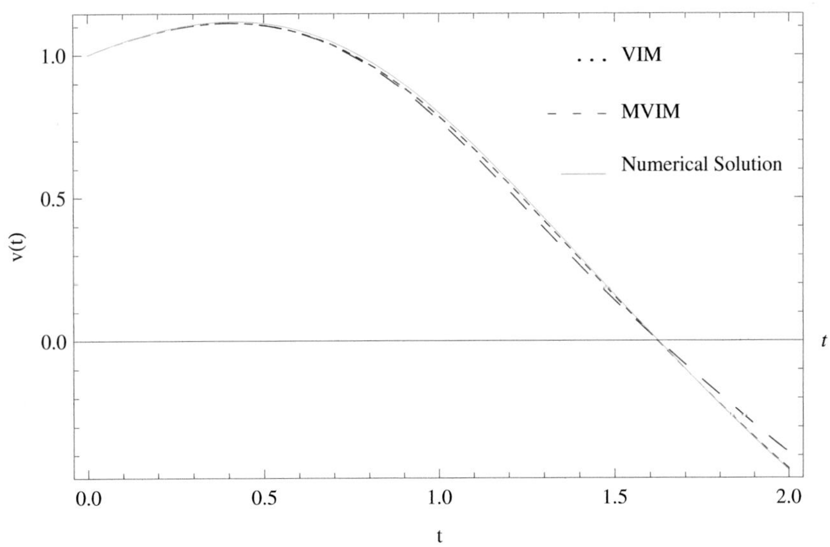

Figure 3.7 Simulation of the approximate results for $v(t) = v_4(t)$ and $\alpha = 0.98$ for Example 3.5.

CHAPTER 4

Differential Transform Method for Fractional Differential Equations

DIFFERENTIAL TRANSFORM METHOD (DTM) is a semi-analytical numerical technique that formulizes Taylor series for the solution of ordinary differential equations, integro-differential equations and partial differential equations. The concept of the differential transform was first proposed by Zhou [116] in a study about electrical circuits, since then it has been applied to many practical applications. The DTM gives exact values of the nth derivative of an analytic function at a point in terms of known and unknown boundary conditions in a fast manner. This method constructs, for differential equations, an analytical solution in the form of a polynomial. Not like the traditional high-order Taylor series method that requires symbolic computation, the DTM is an iterative procedure for obtaining Taylor series solutions. The important advantage is that this method reduces the size of computational work while the Taylor series method is computationally taken a long time for large orders. This transformation technique has been extended to solve fractional differential equations (FDE) by the formulization of the fractional power series in the study of Arikoglu and Ozkol [7] that will be named as fractional differential transform method (FDTM). The formulization in Ref. [7] depends on Caputo fractional derivative and is capable of solving a wide range of FDEs being linear or nonlinear. Also this method applies to any order of fractional differential equations and fractional partial differential equations having homogenous or non-homogenous initial conditions and is named as fractional differential transform method (FDTM) [38, 56].

4.1 FRACTIONAL DIFFERENTIAL TRANSFORM METHOD

In this section, we introduce the fractional differential transform method to obtain approximate analytical solutions for fractional differential equations [7]. Let us

DOI: 10.1201/9781003474609-4

expand the analytical and continuous function $f(x)$ in terms of a fractional power series as follows:

$$f(x) = \sum_{k=0}^{\infty} F(k)(x - x_0)^{\frac{k}{\alpha}}, \tag{4.1}$$

where α is the order of fraction and $F(k)$ is the fractional differential transform of $f(x)$. In order to avoid fractional initial and boundary conditions, according to the relation between the Riemann-Liouville operator and Caputo operator as

$$D^{\beta} f(x) = {}_{x_0}\mathbb{D}_x^{\beta} f(x) \left[f(x) - \sum_{k=0}^{n-1} \frac{1}{k!}(x - x_0)^k f^{(k)}(x_0) \right]. \tag{4.2}$$

Setting $f(x) = f(x) - \sum\limits_{k=0}^{n-1} \frac{1}{k!}(x - x_0)^k f^{(k)}(x_0)$ in Riemann-Liouville fractional derivative and using Eq. (4.2), we obtain fractional derivative in the Caputo sense as follows:

$$D^{\beta} f(x) = \frac{1}{\Gamma(n-\beta)} \frac{d^n}{dx^n} \left\{ \int_{x_0}^{x} \left[\frac{f(t) - \sum_{k=0}^{n-1} \frac{1}{k!}(x - x_0)^k f^{(k)}(x_0)}{(x-t)^{\beta+1-n}} dt \right] \right\}. \tag{4.3}$$

Since the initial conditions are implemented to the integer order derivatives, the transformation of the initial conditions are defined as follows:

$$F(k) = \begin{cases} \text{If } \frac{k}{\alpha} \in Z^+, \frac{1}{(\frac{k}{\alpha})!} \left[\frac{d^{\frac{k}{\alpha}} f(x)}{dx^{\frac{k}{\alpha}}} \right]_{x=x_0}, & \text{for } k = 0, 1, 2, \ldots, (\beta\alpha - 1), \\ \text{If } \frac{k}{\alpha} \notin Z^+, & 0, \end{cases} \tag{4.4}$$

where β is the order of fractional differential equation considered. Using Eq. (4.1), the theorems of FDTM are introduced below. For proofs and details see [7]:

Theorem 4.1 *If $f(x) = g(x) \pm h(x)$, then $F(k) = G(k) \pm H(k)$.*

Theorem 4.2 *If $f(x) = g(x)h(x)$, then $F(k) = \sum_{l=0}^{k} G(l)H(k-l)$.*

Theorem 4.3 *If $f(x) = g_1(x)g_2(x) \cdots g_{n-1}(x)g_n(x)$, then*

$$\begin{aligned} F(k) &= \sum_{k_{n-1}=0}^{k} \sum_{k_{n-2}=0}^{k_{n-1}} \cdots \sum_{k_2=0}^{k_3} \sum_{k_1=0}^{k_2} G_1(k_1)G_2(k_2 - k_1) \cdots G_{n-1}(k_{n-1} - k_{n-2}) \\ &\quad \times G_n(k - k_{n-1}). \end{aligned}$$

Theorem 4.4 *If $f(x) = (x - x_0)^p$, then $F(k) = \delta(k - \alpha p)$ where*

$$\delta(k) = \begin{cases} 1, & k = 0, \\ 0, & k \neq 0. \end{cases}$$

Theorem 4.5 *If* $f(x) = {}_{x_0}D_x^{\beta}[g(x)]$, *then* $F(k) = \dfrac{\Gamma(\beta+1+\frac{k}{\alpha})}{\Gamma(1+\frac{k}{\alpha})}G(k+\alpha\beta)$.

Theorem 4.6 *If* $f(x) = \frac{d^{\beta_1}}{dx^{\beta_1}}[g_1(x)]\frac{d^{\beta_2}}{dx^{\beta_2}}[g_2(x)]\cdots\frac{d^{\beta_{n-1}}}{dx^{\beta_{n-1}}}[g_{n-1}(x)]\frac{d^{\beta_n}}{dx^{\beta_n}}[g_n(x)]$, *then*

$$\begin{aligned}
F(k) &= \sum_{k_{n-1}=0}^{k}\sum_{k_{n-2}=0}^{k_{n-1}}\cdots\sum_{k_2=0}^{k_3}\sum_{k_1=0}^{k_2}\frac{\Gamma\left(\beta_1+1+\frac{k_1}{\alpha}\right)}{\Gamma\left(1+\frac{k_1}{\alpha}\right)}\times\frac{\Gamma\left(\beta_2+1+\frac{k_2-k_1}{\alpha}\right)}{\Gamma\left(1+\frac{k_2-k_1}{\alpha}\right)} \\
&\quad\cdots\frac{\Gamma(\beta_{n-1}+1+\frac{k_{n-1}-k_{n-2}}{\alpha})}{\Gamma(1+\frac{k_{n-1}-k_{n-2}}{\alpha})}\times\frac{\Gamma(\beta_n+1+\frac{k-k_{n-1}}{\alpha})}{\Gamma(1+\frac{k-k_{n-1}}{\alpha})}\times G_1(k_1+\alpha\beta_1) \\
&\quad\times G_2(k_2-k_1+\alpha\beta_2)\cdots G_{n-1}(k_{n-1}-k_{n-2}+\alpha\beta_{n-1}) \\
&\quad\times G_{n-1}(k-k_{n-1}+\alpha\beta_n),
\end{aligned}$$

4.1.1 Applications

Example 4.1 *We consider the linear fractional differential equation [7]*

$$D^{0.5}y(t) = -y(t) + x^2 + \frac{2}{\Gamma(2.5)}x^{1.5}, \tag{4.5}$$

with the condition

$$y(0) = 0. \tag{4.6}$$

Taking $\alpha = 2$ and using Theorems 4.4 and 4.5, Eq. (4.5) transforms to

$$Y(k+1) = \frac{\Gamma\left(1+\frac{k}{2}\right)}{\Gamma\left(\frac{3}{2}+\frac{k}{2}\right)}\left[\delta(k-4) - Y(k) + \frac{2}{\Gamma(2.5)}\delta(k-3)\right], \tag{4.7}$$

and Eq. (4.6) transforms to

$$Y(0) = 0. \tag{4.8}$$

Using Eqs. (4.7) and (4.8), we have

$$\begin{aligned}
k=0,\quad Y(1) &= \frac{\Gamma(1)}{\Gamma(\frac{3}{2})}[-Y(0)] = 0, \\
k=1,\quad Y(2) &= \frac{\Gamma(\frac{3}{2})}{\Gamma(2)}[-Y(1)] = 0, \\
k=2,\quad Y(3) &= \frac{\Gamma(2)}{\Gamma(\frac{5}{2})}[-Y(2)] = 0, \\
k=3,\quad Y(4) &= \frac{\Gamma(\frac{5}{2})}{\Gamma(3)}\left[-Y(3) + \frac{2}{\Gamma(\frac{5}{2})}\right] = 1, \\
k=4,\quad Y(5) &= \frac{\Gamma(3)}{\Gamma(\frac{7}{2})}[-Y(4)+1] = 0,
\end{aligned}$$

and

$$Y(k) = 0, \quad k = 5, 6, \cdots,$$

then using Eq. (4.1), $y(x)$ is evaluated as

$$y(t) = t^2 \tag{4.9}$$

which is the exact solution to the problem.

Example 4.2 *We consider the Bagley-Torvik equation [7].*

$$D^2y(t) + D^{\frac{3}{2}}y(t) + y(t) = 1 + t, \tag{4.10}$$

with the following initial conditions:

$$y(0) = 1, \quad y'(0) = 1. \tag{4.11}$$

Selecting the order of fraction as $\alpha = 2$, the initial conditions are transformed by using Eq. (4.4) as follows:

$$Y(0) = 1, \quad y(1) = 0, \quad Y(2) = 1, \quad Y(3) = 0, \tag{4.12}$$

By using Theorems 4.4 and 4.5, the transform of Eq. (4.10) leads to the following recurrence relation:

$$Y(k+4) = -\frac{\Gamma(\frac{5}{2} + \frac{k}{2})Y(k+3) + \Gamma(1 + \frac{k}{2})[Y(k) - \delta(k-2) - \delta(k)]}{\Gamma(3 + \frac{k}{2})}. \tag{4.13}$$

Using Eqs. (4.12) and (4.13) $Y(k)$ is evaluated up to a certain number of terms and then using the inverse transformation rule, $y(t)$ is evaluated as follows.

$$y(t) = t + 1. \tag{4.14}$$

that Eq. (4.14) is the exact solution.

Example 4.3 *We consider the composite fractional oscillation equation [7]*

$$D^2y(t) - aD^{\beta}y(t) - by(t) = 8, \quad t > 0, \quad 0 < \alpha \leq 2, \tag{4.15}$$

with the initial conditions

$$y(0) = 0, \quad y'(0) = 0. \tag{4.16}$$

Taking $a = b = -1$ and using Theorems 4.4 and 4.5, Eq. (4.15) can be transformed as follows:

$$Y(k+2\alpha) = -\frac{\Gamma\left(\beta + 1 + \frac{k}{\alpha}\right)U(k + \alpha\beta) + \Gamma\left(1 + \frac{k}{\alpha}\right)\left[U(k) - 8\delta(k)\right]}{\Gamma\left(3 + \frac{k}{\alpha}\right)}, \tag{4.17}$$

where α is the unknown value of the fraction. The conditions in Eq. (4.16) can be transformed by using Eq. (4.4) as follows:

$$Y(k) = 0, \quad k = 0, 1, 2, \ldots, 2\alpha - 1. \tag{4.18}$$

Using Eqs. (4.17) and (4.18), $Y(k)$ for $k = 2\alpha, 2\alpha + 1, \ldots, n$ is calculated and using the inverse transformation rule in Eq. (4.1), $y(t)$ is calculated for different values of β. Comparison of numerical results using FDTM $N = 30$ terms, Adomian decomposition method [82] and exact solution are given in Table 4.1, where the exact solution y_{exact} refers to the closed form series solution presented in Ref. [82].

Example 4.4 *Here, we consider the following system of nonlinear fractional differential equations [29]:*

$$\begin{cases} D^{1.3} y_1 = y_1 + y_2^2, \\ D^{2.4} y_2 = y_1 + 5y_2, \end{cases} \tag{4.19}$$

with the initial conditions

$$y_1(0) = 0, \quad y_1'(0) = 1, \quad y_2(0) = 0, \quad y_2'(0) = 1, \quad y_1''(0) = 1. \tag{4.20}$$

By using Theorems 4.1, 4.2 and 4.5, system (4.19) transforms to

$$\begin{aligned} Y_1(k+13) &= \frac{\Gamma(1 + \frac{k}{10})}{\Gamma(1.3 + 1 + \frac{k}{10})} \left[Y_1(k) + \sum_{k_1=0}^{k} Y_2(k_1) Y_2(k - k_1) \right], \\ Y_2(k+24) &= \frac{\Gamma(1 + \frac{k}{10})}{\Gamma(2.4 + 1 + \frac{k}{10})} [Y_1(k) + 5Y_2(k)]. \end{aligned} \tag{4.21}$$

Using Eq. (4.4) the initial conditions in Eq. (4.20) are transformed to

$$\begin{aligned} & Y_1(k) = 0, \qquad k = 0, 1, \ldots, 9, 11, 12, \\ & Y_1(10) = 1, \end{aligned}$$

TABLE 4.1 Comparison of Numerical Results with FDTM and ADM for Example 4.3

t	$\beta = 0.5$			$\beta = 1.5$		
	u_{ADM}	u_{FDTM}	u_{exact}	u_{ADM}	u_{FDTM}	u_{exact}
0.1	0.03974	0.03975	0.03975	0.036478	0.033507	0.033507
0.2	0.158512	0.157036	0.157036	0.140640	0.125221	0.125221
0.3	0.353625	0.347370	0.347370	0.307485	0.267609	0.267609
0.4	0.622083	0.604695	0.604695	0.533284	0.455435	0.455435
0.5	0.960047	0.921768	0.921768	0.814757	0.684335	0.684335
0.6	1.363093	1.290457	1.290457	1.148840	0.950393	0.950393
0.7	1.826257	1.702008	1.702008	1.532571	1.249959	1.249959
0.8	2.344224	2.147287	2.147287	1.963033	1.579557	1.579557
0.9	2.911278	2.617001	2.617001	2.437331	1.935832	1.935832
1.0	3.521462	3.101906	3.101906	2.952567	2.315526	2.315526

$$Y_2(k) = 0, \qquad k = 0, \ldots, 9, 11, \ldots, 19, 21, 22, 23, \tag{4.22}$$
$$Y_2(10) = 1, \quad Y_2(20) = \frac{1}{2}.$$

Using Eqs. (4.21) and (4.22), $Y_1(k)$ is obtained

$$\begin{aligned}
&k = 0, 1, \ldots, 9, \quad Y_1(k+13) = 0,\\
&k = 10, \quad Y_1(23) = \frac{1}{\Gamma\left(\frac{33}{10}\right)},\\
&k = 11, \ldots, 19 \quad Y_1(k+13) = 0,\\
&k = 20, \quad Y_1(33) = \frac{2}{\Gamma\left(\frac{43}{10}\right)},\\
&k = 21, 22 \quad Y_1(34) = Y_1(35) = 0,\\
&k = 23, \quad Y_1(36) = \frac{1}{\Gamma\left(\frac{23}{5}\right)},\\
&\vdots
\end{aligned} \tag{4.23}$$

and $Y_2(k)$ is obtained

$$\begin{aligned}
&k = 0, 1, \ldots, 9 \quad Y_2(k+24) = 0,\\
&k = 10, \quad Y_2(34) = \frac{6}{\Gamma\left(\frac{22}{5}\right)},\\
&k = 11, \ldots, 19 \quad Y_2(k+24) = 0,\\
&k = 20, \quad Y_2(44) = \frac{5}{\Gamma\left(\frac{27}{5}\right)},\\
&k = 21, 22 \quad Y_2(45) = Y_2(46) = 0,\\
&k = 23, \quad Y_2(47) = \frac{1}{\Gamma\left(\frac{57}{10}\right)},\\
&\vdots
\end{aligned} \tag{4.24}$$

then using Eq. (4.1), the following series solutions are obtained for $y_1(t)$ and $y_2(t)$, respectively:

$$\begin{aligned}
y_1(t) &= t + \frac{t^{\frac{23}{10}}}{\Gamma\left(\frac{33}{10}\right)} + \frac{2t^{\frac{33}{10}}}{\Gamma\left(\frac{43}{10}\right)} + \frac{t^{\frac{18}{5}}}{\Gamma\left(\frac{23}{5}\right)} + \frac{6t^{\frac{43}{10}}}{\Gamma\left(\frac{53}{10}\right)} + \frac{2t^{\frac{23}{5}}}{\Gamma\left(\frac{28}{5}\right)} + \frac{t^{\frac{49}{10}}}{\Gamma\left(\frac{59}{10}\right)} + \cdots,\\
y_2(t) &= t + \frac{t^2}{2} + \frac{6t^{\frac{17}{15}}}{\Gamma\left(\frac{22}{5}\right)} + \frac{5t^{\frac{22}{5}}}{\Gamma\left(\frac{27}{5}\right)} + \frac{t^{\frac{47}{10}}}{\Gamma\left(\frac{57}{10}\right)} + \cdots.
\end{aligned} \tag{4.25}$$

In Figure 4.1, we draw the curves $y_1(t)$ and $y_2(t)$. The graphical results are in very good agreement with the results in Ref. [20].

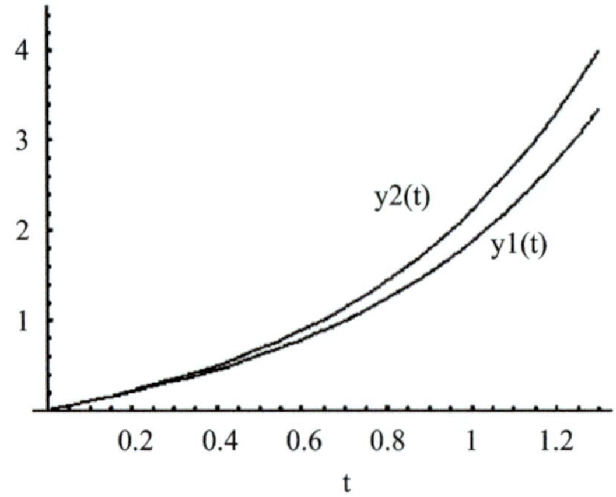

Figure 4.1 Approximate solutions of y_1 and y_2 with DTM for Example 4.4.

4.2 FRACTIONAL REDUCED DIFFERENTIAL TRANSFORM METHOD

In this section, we introduce the fractional reduced differential transform method (FRDTM) used to obtain approximate analytical solutions for the fractional differential equation. This method has been developed in Refs. [93, 104] as follows.

Let $u(x,t)$ be an analytic function that continuously differentiable with respect to t and space x in domain of interest. Define

$$U_k(x) = \frac{1}{\Gamma(k\alpha+1)} \left[D_t^{k\alpha} u(x,t)\right]_{t=0}, \tag{4.26}$$

where α is a parameter describing the order of time-fractional derivative in the Caputo sense and t-dimensional spectrum function $U_k(x)$ is defined as

$$u(x,t) = \sum_{k=0}^{\infty} U_k(x) t^{k\alpha}. \tag{4.27}$$

Combining equations (4.26) and (4.27), we have

$$u(x,t) = \sum_{k=0}^{\infty} \frac{1}{\Gamma(k\alpha+1)} \left[D_t^{k\alpha} u(x,t)\right]_{t=0} t^{k\alpha}. \tag{4.28}$$

If $\alpha = 1$, fractional reduced differential transformation is converted to reduced differential transformation. Some theorems of the fractional reduced differential transformation, obtained from Eqs. (4.26) and (4.27), are given in below. For proofs and details see [104]:

Theorem 4.7 *If* $w(x,t) = u(x,t) + v(x,t)$*, then* $W_k(x) = U_k(x) \pm V_k(x)$.

Theorem 4.8 *If* $w(x,t) = cu(x,t)$, *then* $W_k(x) = cU_k(x)$, *c is a constant.*

Theorem 4.9 *If* $w(x,t) = x^m t^n$, *then* $W_k(x) = x^m \delta(k\alpha - n)$ *where,*

$$\delta(k) = \begin{cases} 1 & k = 0, \\ 0 & k \neq 0. \end{cases}$$

Theorem 4.10 *If* $w(x,t) = x^m t^n u(x,t)$, *then* $W_k(x) = x^m U_{k-n}(x)$.

Theorem 4.11 *If* $w(x,t) = u(x,t)v(x,t)$, *then*

$$W_k(x) = \sum_{r=0}^{k} U_r(x) V_{k-r}(x) = \sum_{r=0}^{k} U_{k-r}(x) U_r(x).$$

Theorem 4.12 *If* $w(x,t) = u_1(x,t)u_2(x,t)\dots u_{n-1}(x,t)u_n(x,t)$, *then*

$$\begin{aligned} W_k(x) &= \sum_{k_{n-1}=0}^{k} \sum_{k_{n-2}=0}^{k_{n-1}} \cdots \sum_{k_2=0}^{k_3} \sum_{k_1=0}^{k_2} U_{1,k_1}(x) U_{2,k_2-k_1}(x) \dots U_{n-1,k_{n-1}-k_{n-2}}(x) \\ &\quad \times U_{n,k-k_{n-1}}(x). \end{aligned}$$

Theorem 4.13 *If* $w(x,t) = \frac{\partial^m}{\partial x^m} u(x,t)$, *then* $W_k(x) = \frac{\partial^m}{\partial x^m} U_k(x)$.

Theorem 4.14 *If* $w(x,t) = \frac{\partial^r}{\partial t^r} u(x,t)$, *then*

$$W_k(x) = (k+1)(k+2)\dots(k+r)U_{k+r}(x) = \frac{(k+r)!}{r!} U_{k+r}(x).$$

Theorem 4.15 *If* $w(x,t) = D_t^{N\alpha} u(x,t)$, *then* $W_k(x) = \frac{\Gamma(k\alpha+N\alpha+1)}{\Gamma(k\alpha+1)} U_{k+N}(x)$.

4.2.1 Applications

Example 4.5 *We consider the following one-dimensional fractional heat-like problem:*

$$D_t^{\alpha} u = \frac{x^2}{2} \frac{\partial^2 u}{\partial x^2}, \quad 0 < \alpha \leq 1, \tag{4.29}$$

with the initial condition

$$u(x,0) = x^2. \tag{4.30}$$

Applying FRDTM to Eq. (4.29), the recurrence equation is obtained as follows:

$$\frac{\Gamma(\alpha(k+1)+1)}{\Gamma(\alpha k+1)} U_{k+1}(x) = \frac{x^2}{2} \frac{\partial^2 U_k}{\partial x^2}. \tag{4.31}$$

Using FRDTM to the initial conditions (4.30), we get

$$U_0(x) = x^2. \tag{4.32}$$

Using Eqs. (4.32) and (4.31), we get

$$\begin{aligned}
&U_1(x) = \frac{x^2}{\Gamma(\alpha+1)}, \quad U_2(x) = \frac{x^2}{\Gamma(2\alpha+1)},\\
&U_3(x) = \frac{x^2}{\Gamma(3\alpha+1)}, \cdots U_k(x) = \frac{x^2}{\Gamma(k\alpha+1)},\\
&\vdots
\end{aligned} \tag{4.33}$$

By Substituting the values of Eq. (4.33) in (4.31) and using Eq. (4.27) we get

$$\begin{aligned}
u(x,t) &= u(x,t) = \sum_{k=0}^{\infty} U_k(x) t^{k\alpha}\\
&= \sum_{k=0}^{\infty} \frac{x^2}{\Gamma(k\alpha+1)} t^{k\alpha} = x^2 \sum_{k=0}^{\infty} \frac{t^{k\alpha}}{\Gamma(k\alpha+1)} = x^2 E_\alpha(t^\alpha).
\end{aligned}$$

Example 4.6 *Consider the following time fractional-order linear Telegraph equation [104]*

$$D_t^{2\alpha} u + 2D_t^{\alpha} u + u = \frac{\partial^2 u}{\partial x^2}, \tag{4.34}$$

with the initial conditions

$$\begin{cases} u(x,0) = \sinh x,\\ u_t(x,0) = -2\sinh x. \end{cases} \tag{4.35}$$

Applying RDTM to Eq. (4.34), the recurrence equation is obtained as follows:

$$\frac{\Gamma(k\alpha+2\alpha+1)}{\Gamma(k\alpha+1)} U_{k+2}(x) + 2\frac{\Gamma(k\alpha+\alpha+1)}{\Gamma(k\alpha+1)} U_{k+1}(x) = \frac{\partial^2}{\partial x^2} U_k(x) - U_k(x). \tag{4.36}$$

Using FRDTM to the initial conditions (4.35), we get

$$U_0(x) = \sinh x, \qquad U_1(x) = -2\sinh x. \tag{4.37}$$

Using Eq. (4.37) in (4.36), we get the following $U_k(x)$ values as

$$U_k(x) = \frac{(-2)^k}{\Gamma\left(\frac{k}{\lambda}+1\right)} \Gamma\left(\frac{\lambda+1}{\lambda}\right) \sinh x, \quad k \geq 2, \tag{4.38}$$

where $\alpha = \frac{1}{\lambda}, \lambda > 0$. Using the differential inverse reduced transform of $U_k(x)$, we get

$$
\begin{aligned}
u(x,t) &= \sum_{k=0}^{\infty} U_k(x)t^{k\alpha} = \sum_{k=0}^{\infty} U_k(x)t^{\frac{k}{\lambda}} \\
&= U_0(x) + U_1(x)t^{\frac{1}{\lambda}} + U_2(x)t^{\frac{2}{\lambda}} + U_3(x)t^{\frac{1}{\lambda}} + \cdots \\
&= \sinh x \left[1 + (-2)t^{\frac{1}{\lambda}} + \Gamma\left(\frac{\lambda+1}{\lambda}\right)\left\{\frac{(-2)^2}{\Gamma\left(\frac{2}{\lambda}+1\right)}t^{\frac{2}{\lambda}} + \frac{(-2)^3}{\Gamma\left(\frac{3}{\lambda}+1\right)}t^{\frac{3}{\lambda}} + \cdots\right\}\right].
\end{aligned}
\tag{4.39}
$$

When $\lambda = 1$, we get the exact solution of the non-fractional form of the Telegraph equation (4.34) as

$$
\begin{aligned}
u(x,t) &= \sinh x \left(1 + (-2)t + \frac{(-2)^2}{2!}t^2 + \frac{(-2)^3}{3!}t^3 + \cdots + \frac{(-2)^k}{k!}t^k + \cdots\right) \\
&= e^{-2t}\sinh x.
\end{aligned}
\tag{4.40}
$$

Example 4.7 *Consider the following time fractional-order nonlinear Telegraph equation [104]*

$$
D_t^{2\alpha}u + 2D_t^{\alpha}u + u^2 = \frac{\partial^2 u}{\partial x^2} + e^{(2x-4t)} - e^{(x-2t)}, \tag{4.41}
$$

under the initial conditions

$$
\begin{cases} u(x,0) = e^x, \\ u_t(x,0) = -2e^x. \end{cases} \tag{4.42}
$$

Applying FRDTM to Eq. (4.41), the recurrence equation is obtained as follows:

$$
\frac{\Gamma(k\alpha+2\alpha+1)}{\Gamma(k\alpha+1)}U_{k+2}(x) + 2\frac{\Gamma(k\alpha+\alpha+1)}{\Gamma(k\alpha+1)}U_{k+1}(x) \tag{4.43}
$$

$$
= \frac{\partial^2}{\partial x^2}U_k(x) - \sum_{r=0}^{k} U_r(x)U_{k-r}(x) + e^{2x}\left(\frac{(-4)^k}{k!}\right) - e^x\left(\frac{(-2)^k}{k!}\right). \tag{4.44}
$$

Using FRDTM to the initial conditions (4.42), we get

$$
U_0(x) = e^x \qquad U_1(x) = -2e^x. \tag{4.45}
$$

Using Eq. (4.37) in (4.36), we get the following $U_k(x)$ values as

$$
U_2(x) = 4\left(\frac{\Gamma(\alpha+1)}{\Gamma(2\alpha+1)}\right)e^x, \tag{4.46}
$$

$$
U_3(x) = -8\left(\frac{\Gamma(\alpha+1)}{\Gamma(3\alpha+1)}\right)e^x, \tag{4.47}
$$

$$
\vdots \tag{4.48}
$$

Using the differential inverse reduced transform of $U_k(x)$, we get

$$\begin{aligned} u(x,t) &= \sum_{k=0}^{\infty} U_k(x) t^{k\alpha} \quad (4.49) \\ &= e^x \left[1 + (-2)t^{\alpha} + 4\left(\frac{\Gamma(\alpha+1)}{\Gamma(2\alpha+1)}\right) t^{2\alpha} + 8\left(-\frac{\Gamma(\alpha+1)}{\Gamma(3\alpha+1)}\right) t^{3\alpha} + \cdots \right]. \end{aligned}$$

When $\alpha = 1$, the exact solution of the non-fractional form of the nonlinear Telegraph equation (4.41) is obtained

$$u(x,t) = e^{x-2t}. \quad (4.50)$$

CHAPTER 5

Numerical Methods for Fractional Differential Equations

FOR SCIENTITIC APPLICATIONS, it is often necessary to solve fractional differential equations. Most fractional differential equations for practical problems cannot be solved analytically. Therefore, numerical methods for fractional differential equations are extremely important. Almost all numerical methods for fractional differential equations are based on the approximation of fractional differential or integral operators by appropriate formulas. In this chapter, we study numerical methods for fractional differential equations such as Adams-Bashforth-Moulton method, Diethelm's method and Grünwald-Letnikov's method.

5.1 ADAMS-BASHFORTH-MOULTON METHOD

In this section, we apply the fundamental algorithm to the solution of initial value problems with Caputo derivatives. The algorithm is a generalization of the classical Adams-Bashforth-Moulton method that is well known for the numerical solution of first-order problems. This method is introduced by Diethelm and Freed [22, 23]. It is based on the approximation of the integral operator by the product trapezoidal method.

We first consider the fractional ordinary differential equation described as follows [21]:

$$D^{\alpha}y(t) = f(t, y(t)), \qquad m-1 < \alpha \leq m, \quad 0 \leq t \leq T, \tag{5.1}$$

along with initial conditions

$$y^{(k)}(0) = y_0^{(k)}, \qquad k = 0, 1, \ldots, m-1. \tag{5.2}$$

By applying Riemann-Liouville integral operator I^{α} to Eq. (5.1), the initial value problem converts to an equivalent Volterra integral equation,

$$y(t) = \sum_{k=0}^{m-1} \frac{y_0^{(k)}}{k!} t^k + I^{\alpha}[f(t, y(t))] \tag{5.3}$$

DOI: 10.1201/9781003474609-5

that we have the following formula

$$y(t) = \sum_{k=0}^{m-1} y_0^{(k)} \frac{t^k}{k!} + \frac{1}{\Gamma(\alpha)} \int_0^t (t-\tau)^{\alpha-1} f(\tau, y(\tau)) d\tau. \tag{5.4}$$

that a continuous function is a solution of the initial value problem if and only if it is a solution of Eq. (5.4).

We simply use the product trapezoidal quadrature formula to replace the integral, where nodes $t_j (j = 0, 1, \ldots, n+1)$ are taken with respect to the weight function $(t_{n+1} - \tau)^{\alpha-1}$. In other words, we apply the approximation

$$\int_0^{t_{n+1}} (t_{n+1} - \tau)^{\alpha-1} g(\tau) d\tau \simeq \int_0^{t_{n+1}} (t_{n+1} - \tau)^{\alpha-1} \tilde{g}_{n+1}(\tau) d\tau, \tag{5.5}$$

where $\tilde{g}_{n+1}$ is the piecewise linear interpolant for g with nodes and knots chosen at the $t_j, j = 0, 1, 2, \ldots, n+1$. Using standard techniques from quadrature theory [23], we find that we can write the integral on the right-hand side of Eq. (5.5) as

$$\int_0^{t_{n+1}} (t_{n+1} - \tau)^{\alpha-1} \tilde{g}_{n+1}(\tau) d\tau = \frac{h^\alpha}{\alpha(\alpha+1)} \sum_{j=0}^{n+1} a_{j,n+1} g(t_j),$$

where

$$a_{j,n+1} = \begin{cases} n^{\alpha+1} - (n-\alpha)(n+1)^\alpha, & \text{if } j = 0, \\ (n-j+2)^{\alpha+1} + (n-j)^{\alpha+1} - 2(n-j+1)^{\alpha+1}, & \text{if } 1 \le j \le n. \\ 1, & \text{if } j = n+1. \end{cases} \tag{5.6}$$

This then gives us our corrector formula (i.e., the fractional variant of the one-step Adams-Moulton method), which is

$$\begin{aligned} y(t_{n+1}) &= \sum_{k=0}^{m-1} y_0^{(k)} \frac{t_{n+1}^k}{k!} + \frac{h^\alpha}{\Gamma(\alpha+2)} f\left[t_{n+1}, y^p(t_{n+1})\right] \\ &\quad + \frac{h^\alpha}{\Gamma(\alpha+2)} \sum_{j=0}^{n} a_{j,n+1} f\left(t_j, y_n(t_j)\right). \end{aligned} \tag{5.7}$$

where we have used the functional equation of the Gamma function twice and the fact that $a_{n+1,n+1} = 1$.

The remaining problem is the determination of the predictor formula that we require to calculate the value $y^p(t_{n+1})$. We replace the integral on the right-hand side of Eq. (5.4) by the product rectangle rule

$$\int_0^{t_{n+1}} (t_{n+1} - \tau)^{\alpha-1} g_{n+1}(\tau) d\tau \approx \sum_{j=0}^{n} b_{j,n+1} g(t_j),$$

where

$$b_{j,n+1} = \frac{h^\alpha}{\alpha}\left[(n+1-j)^\alpha - (n-j)^\alpha\right]. \tag{5.8}$$

Thus, the predicted value $y^p(t_{n+1})$ is determined by the fractional Adams Bashforth method

$$y^p(t_{n+1}) = \sum_{k=0}^{m-1} \frac{t_{n+1}^k}{k!} y_0^{(k)} + \frac{1}{\Gamma(\alpha)} \sum_{j=0}^{n} b_{j,n+1} f\left(t_j, y_n(t_j)\right). \tag{5.9}$$

Our basic algorithm, the fractional Adams-Bashforth-Moulton method, is completely described now by Eqs. (5.7) and (5.9) with the weights $a_{j,n+1}$ and $b_{j,n+1}$ being defined according to Eqs. (5.6) and (5.8), respectively.

5.1.1 Applications

Example 5.1 *Consider the following fractional equation*

$$D^\alpha y(t) = y^2(t) + 1, \quad 0 < t < 1, \quad 0 < \alpha < 1, \tag{5.10}$$

with the initial condition,

$$y(0) = 0, \tag{5.11}$$

The exact solution of this problem when $\alpha = 1$ is

$$y(t) = \tan t.$$

We have taken the step size $h = 0.1$, 0.01 and $h = 0.001$ in this example. The results of our computation agree with the analytical solution. Figure 5.1 shows exact solution and approximate solution obtained $y(t)$ of Eq. (5.10) for $h = 0.01$ and different values of α. As illustrated in Figure 5.1, α approaches 1 causes the numerical solution to converges to the exact solution. Table 5.1 shows the absolute errors of $y(t)$ for $\alpha = 1$ and the different values of h. It is obvious that the numerical solution to converge to the exact solution when the step size h becomes small. It is apparent from these results that the suggested method has satisfactory efficiency and accuracy with small terms.

Example 5.2 *Consider the following nonlinear fractional differential equation*

$$D^{1.5} y(t) + e^{-2\pi} y^2(t) = g(t), \quad t \in (0, 1], \tag{5.12}$$

where

$$g(t) = \frac{105\sqrt{\pi}}{32} t^2 + e^{-2\pi} t^7,$$

with the initial condition,

$$y(0) = 0, \quad y'(0) = 0, \tag{5.13}$$

The exact solution of this problem is

$$y(t) = t^{\frac{7}{2}}.$$

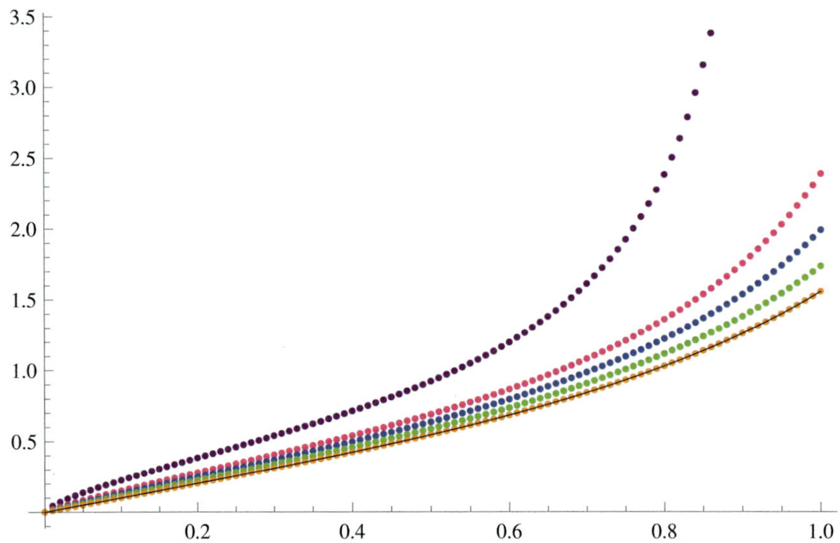

Figure 5.1 Comparison of the exact solution (black solid line) and the approximate solutions for $h = 0.01$ when $\alpha = 1$ (....), $\alpha = 0.95$ (....), $\alpha = 0.9$ (....), $\alpha = 0.85$ (....), $\alpha = 0.5$ (....) for Example 5.1

TABLE 5.1 Absolute Errors of $y(t)$ for $\alpha = 1$ and the Different Values of h for Example 5.1

t	$h = 0.1$	$h = 0.01$	$h = 0.001$
0	0	0	0
0.1	1.65328×10^{-4}	1.64973×10^{-6}	1.64965×10^{-8}
0.2	3.20241×10^{-4}	3.18927×10^{-6}	3.18827×10^{-8}
0.3	4.51186×10^{-4}	4.46134×10^{-6}	4.45607×10^{-8}
0.4	5.35767×10^{-4}	5.19239×10^{-6}	5.1737×10^{-8}
0.5	5.33019×10^{-4}	4.86784×10^{-6}	4.8143×10^{-8}
0.6	3.64621×10^{-4}	2.48064×10^{-6}	2.34419×10^{-8}
0.7	1.24557×10^{-4}	4.01591×10^{-6}	4.34294×10^{-8}
0.8	1.25529×10^{-3}	1.8987×10^{-5}	1.97536×10^{-7}
0.9	3.72992×10^{-3}	5.23664×10^{-5}	5.41832×10^{-7}
1	9.20899×10^{-3}	1.28837×10^{-4}	1.33339×10^{-6}

We have taken the step size $h = 0.1$, 0.01 and $h = 0.001$ in this example. The results of our computation agree with the analytical solution. Figure 5.2 illustrates exact solution and the approximation solution with the present method of Eq. (5.12) for $h = 0.01$. Table 5.2 shows the absolute errors of $y(t)$ for the different values of h. It is obvious that the numerical solution to converge to the exact solution when the step size h becomes small.

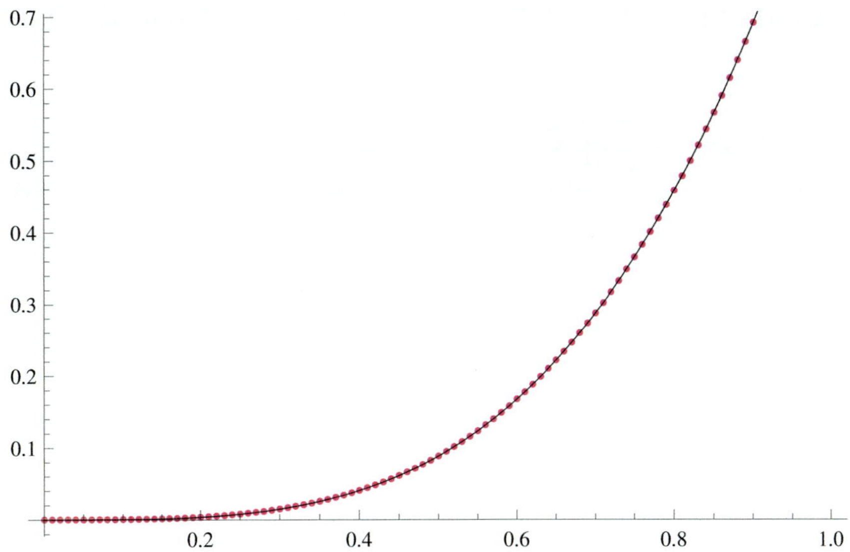

Figure 5.2 Comparison of the exact solution (black line) and approximation solutions with $h = 0.01$ (....) for Example 5.2.

TABLE 5.2 Absolute Errors of $y(t)$ for the Different Values of h for Example 5.2

t	$h = 0.1$	$h = 0.01$	$h = 0.001$
0	0	0	0
0.1	2.37171×10^{-4}	2.30852×10^{-6}	2.30592×10^{-8}
0.2	6.59584×10^{-4}	6.52464×10^{-6}	6.52196×10^{-8}
0.3	1.20591×10^{-3}	1.19843×10^{-5}	1.19815×10^{-7}
0.4	1.85266×10^{-3}	1.84195×10^{-5}	1.84467×10^{-7}
0.5	2.58619×10^{-3}	2.57829×10^{-5}	2.57799×10^{-7}
0.6	3.39728×10^{-3}	3.38917×10^{-5}	3.38884×10^{-7}
0.7	4.27923×10^{-3}	4.27079×10^{-5}	4.27036×10^{-7}
0.8	5.22688×10^{3}	5.21788×10^{-5}	5.21725×10^{-7}
0.9	6.23619×10^{-3}	6.22617×10^{-5}	6.22519×10^{-7}
1	7.30391×10^{-3}	7.29215×10^{-5}	7.29055×10^{-7}

5.2 DIETHELM'S APPROACH

There are of course many variants on how to discretize derivatives, both of integer and of fractional order, see for example Lambert [72] and Lubich [76]. A numerical algorithm for computing Caputo derivatives has been derived by Diethelm [24] and is recalled here. We will need to approximate derivatives of both integer and fractional order.

To discretize the Caputo fractional derivative we will use Diethelm's method [26]:

$$D^\alpha y(t) = \frac{1}{{}^\alpha\gamma_i}\left(\sum_{k=0}^{i} {}^\alpha\omega_{k,i} y_{i-k} + \frac{y_0}{\alpha}\right), \quad i = 1, 2, \ldots, n, \tag{5.14}$$

where ${}^\alpha\gamma_i = (ih)^\alpha \Gamma(-\alpha)$ and ${}^\alpha\omega_{k,0}, \ldots, {}^\alpha\omega_{k,i}$ are convolution weights derived from the fact that the fractional operator is defined in terms of a convolution integral. where the weights $\omega_{k,j}$ are given by the following expressions:

$${}^\alpha\omega_{k,j} = \frac{j^\alpha}{\alpha(1-\alpha)}\begin{cases} -1, & \text{if } k = 0, \\ 2k^{1-\alpha} - (k-1)^{1-\alpha} - (k+1)^{1-\alpha}, & \text{if } k = 1, 2, \ldots, j-1, \\ (\alpha - 1)k^{-\alpha} - (k-1)^{1-\alpha} + k^{1-\alpha}, & \text{if } k = j. \end{cases} \tag{5.15}$$

To solve a first-order differential equation, we will use the trapezium rule:

$$Dy(t) = f \Rightarrow y_i = y_{i-1} + \frac{1}{2}h\left(f_i + f_{i-1}\right). \tag{5.16}$$

It should be noted that D and D^α are operators and so their appearance in matrices is a notational device.

Suppose now that we wish to discretize equation [26]

$$Dy + a_1 D^{\alpha_2} + a_2 D^{\alpha_1} + a_3 y = f, \qquad 1 > \alpha_2 > \alpha_1 > 0, \tag{5.17}$$

we could write Eq. (5.17) as the system

$$\begin{aligned} y_1 &= y, \\ y_2 &= D^{\alpha_1} y_1, \\ y_3 &= D^{\alpha_2} y_1 = D^{\alpha_2 - \alpha_1} y_2, \\ y_4 &= D y_1 = D^{1-\alpha_2} y_3. \end{aligned}$$

We choose the former since it requires the weights for one fewer fractional derivative to be calculated and has less propagation of rounding error.

Consider, for example, the equation

$$D^2 y + a_3 D^{\alpha_3} + a_2 Dy + a_1 D^{\alpha_1} y + a_0 y = f, \quad \alpha_3 = 1 + \beta, \quad \alpha_1 = \alpha, \quad \alpha, \beta \in (0,1), a_i \in \mathbb{R}, \tag{5.18}$$

we write Eq. (5.18) as the system

$$\begin{aligned} y_1 &= y, \\ y_2 &= D^\alpha y = D^{\alpha_1} y_1, \\ y_3 &= Dy = Dy_1, \\ y_4 &= D^\beta D y = D^\beta y_3, \\ y_5 &= D^2 y = D y_3. \end{aligned}$$

which in matrix form is

$$\begin{pmatrix} D^\alpha & 0 & 0 & 0 \\ D & 0 & 0 & 0 \\ 0 & 0 & D^\beta & 0 \\ 0 & 0 & D & 0 \end{pmatrix} \begin{pmatrix} y_1(t) \\ y_2(t) \\ y_3(t) \\ y_4(t) \end{pmatrix} = \begin{pmatrix} y_2(t) \\ y_3(t) \\ y_4(t) \\ f(t) - \sum_{j=1}^4 a_{j-1} y_j(t) \end{pmatrix}.$$

In view of Eq. (5.14), consider the equation $D^\alpha y_1 = y_2$. We set our approximations $y_j \approx y(jh)$ for some fixed $h > 0$ and we discretise this first component as

$$\frac{1}{{}^\alpha\gamma_i}\left(\sum_{k=0}^{i} {}^\alpha\omega_{k,i} y_{1i-k} + \frac{y_{10}}{\alpha}\right) = y_{2i}, \tag{5.19}$$

and put

$$S_{1i} = \sum_{k=1}^{i} {}^\alpha\omega_{k,i} y_{1i-k} + \frac{y_{10}}{\alpha},$$

so we have

$${}^\alpha\omega_{0,i} y_{1i} + S_{1i} = {}^\alpha\gamma_i y_{2i},$$

and it follows that

$$-{}^\alpha\omega_{0,i} y_{1i} + {}^\alpha\gamma_i y_{2i} = S_{1i},$$

Now for the final row, we have

$$y_{3i} = y_{3i-1} + \frac{h}{2}\left(f_i + f_{i-1} - \sum_{j=1}^{4} a_{j-1}(y_{ji} - y_{ji-1})\right),$$

so with $F_i = (f_i + f_{i-1})\frac{h}{2}$ and

$$S_{4i} = y_{3i-1} + F_i - \frac{h}{2}\sum_{j=1}^{4} a_{j-1} y_{ji-1},$$

we have

$$\frac{ha_0}{2} y_{1i} + \frac{ha_1}{2} y_{2i} + \left(1 + \frac{ha_2}{2}\right) y_{3i} + \frac{ha_3}{2} y_{4i} = S_{4i}.$$

This leads to the matrix form:

$$\begin{pmatrix} -{}^\alpha\omega_{0,i} & {}^\alpha\gamma_i & 0 & 0 \\ 1 & 0 & -\frac{h}{2} & 0 \\ 0 & 0 & -{}^\beta\omega_{0,i} & {}^\beta\gamma_i \\ \frac{ha_0}{2} & \frac{ha_1}{2} & 1 + \frac{ha_2}{2} & \frac{ha_3}{2} \end{pmatrix} \begin{pmatrix} y_{1i} \\ y_{2i} \\ y_{3i} \\ y_{4i} \end{pmatrix} = \begin{pmatrix} S_{1i} \\ S_{2i} \\ S_{3i} \\ S_{4i} \end{pmatrix}.$$

where

$$\begin{aligned}
S_{1i} &= \sum_{k=1}^{i} {}^{\alpha}\omega_{k,i} y_{1i-k} + \frac{y_{10}}{\alpha}, \\
S_{2i} &= y_{1i-1} + \frac{h}{2} y_{3i-1}, \\
S_{3i} &= \sum_{k=1}^{i} {}^{\beta}\omega_{k,i} y_{3i-k} + \frac{y_{30}}{\beta}, \\
S_{4i} &= y_{3i-1} + F_i - \frac{h}{2}\sum_{j=1}^{4} a_{j-1} y_{ji-1},
\end{aligned}$$

and $F_i = \frac{h(f_i+f_{i-1})}{2}$, $f_i = f(ih)$.

5.2.1 Applications

Example 5.3 *Consider the following Bagley-Torvik equation [26]*

$$D^2 y(t) + D^{\alpha_1} y(t) + y(t) = 0, \qquad \alpha_1 = 1+\alpha \quad \alpha \in (0,1), \tag{5.20}$$

with the initial condition , $y(0) = 1$, $y^{'}(0) = 0$.

We rewrite Eq. (5.20) as

$$y_1 = y, \qquad y_2 = Dy, \qquad y_3 = D^{\alpha} Dy, \quad y_4 = D^2 y,$$

which written in matrix notation is

$$\begin{pmatrix} D & 0 & 0 \\ 0 & D^{\alpha} & 0 \\ 0 & D & 0 \end{pmatrix} \begin{pmatrix} y_1 \\ y_2 \\ y_3 \end{pmatrix} = \begin{pmatrix} y_2 \\ y_3 \\ -y_3 - y_1 \end{pmatrix}.$$

Discretising as above this results in the matrix system

$$\begin{pmatrix} 1 & -\frac{h}{2} & 0 \\ 0 & -{}^{\alpha}\omega_{0,i} & {}^{\alpha}\gamma_i \\ \frac{h}{2} & 1 & \frac{h}{2} \end{pmatrix} \begin{pmatrix} y_{1i} \\ y_{2i} \\ y_{3i} \end{pmatrix} = \begin{pmatrix} y_{1i-1} + \frac{h}{2} y_{2i-1} \\ \sum_{k=1}^{i} y_{2i-k} + \frac{y_{20}}{\alpha} \\ -\frac{h}{2} y_{1i-1} + y_{2i-1} - \frac{h}{2} y_{3i-1} \end{pmatrix}.$$

Figure 5.3 shows the numerical results for $y(t)$ for $\alpha = 0.25$ and Figure 5.4 shows the numerical results for $y(t)$ for $\alpha = 0.5$.

Example 5.4 *We consider the fractional Riccati equation as Ref. [54]*

$$D^{\alpha} y(t) = -y^2(t) + 1, \quad 0 < t < 1, \quad 0 < \alpha < 1, \tag{5.21}$$

with the initial condition

$$y(0) = 0, \tag{5.22}$$

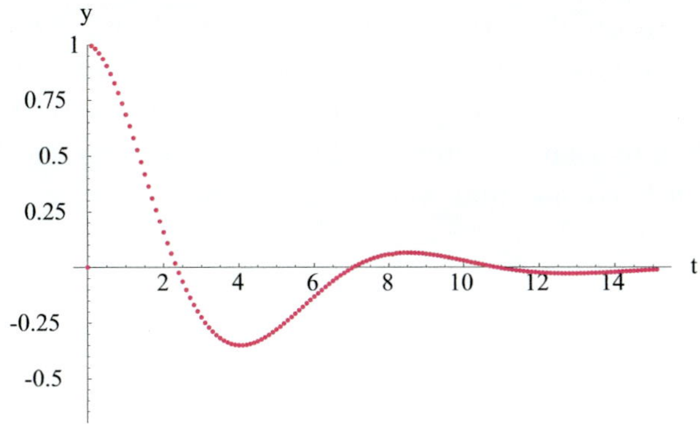

Figure 5.3 Approximate solutions $y(t)$ for $\alpha = 0.25$ for Example 5.3.

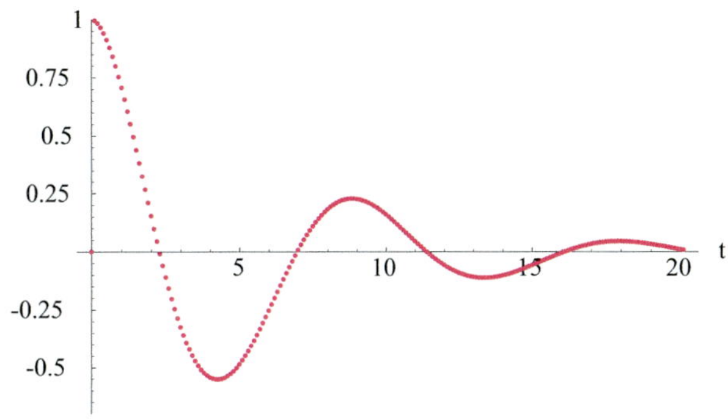

Figure 5.4 Approximate solutions $y(t)$ for $\alpha = 0.5$ for Example 5.3.

the exact solution when $\alpha = 1$ *is*

$$y(t) = \frac{e^{2t} - 1}{e^{2t} + 1}.$$

We discretize the Caputo fractional derivative in Eq. (5.22) as

$$\frac{1}{{}^{\alpha}\gamma_i}\left(\sum_{k=0}^{i} {}^{\alpha}\omega_{k,i} y_{i-k} + \frac{y(0)}{\alpha}\right) = -y_i^2 + 1, \tag{5.23}$$

and put

$$S_i = \left(\sum_{k=1}^{i} {}^{\alpha}\omega_{k,i} y_{i-k} + \frac{y(0)}{\alpha}\right), \tag{5.24}$$

We rewrite Eq. (5.23) in the following form:

$${}^{\alpha}\omega_{0,i} y_i + S_i = -{}^{\alpha}\gamma_i y_i^2 +{}^{\alpha} \gamma_i, \tag{5.25}$$

finally, we have

$$S_i = -{}^{\alpha}\gamma_i y_i^2 +{}^{\alpha} \gamma_i -{}^{\alpha} \omega_{0,i} y_i, \tag{5.26}$$

Figure 5.5 shows the absolute errors of $y(t)$ for $h = 0.01$ and the different values of α. Table 5.3 shows the absolute errors of $y(t)$ for $\alpha = 0.99$ and the different values of h. Also, Table 5.4 shows the absolute errors of $y(t)$ for $h = 0.01$ and the different values of α. It is obvious that the numerical solution converge to the exact solution when the step size h becomes small and α approaches 1.

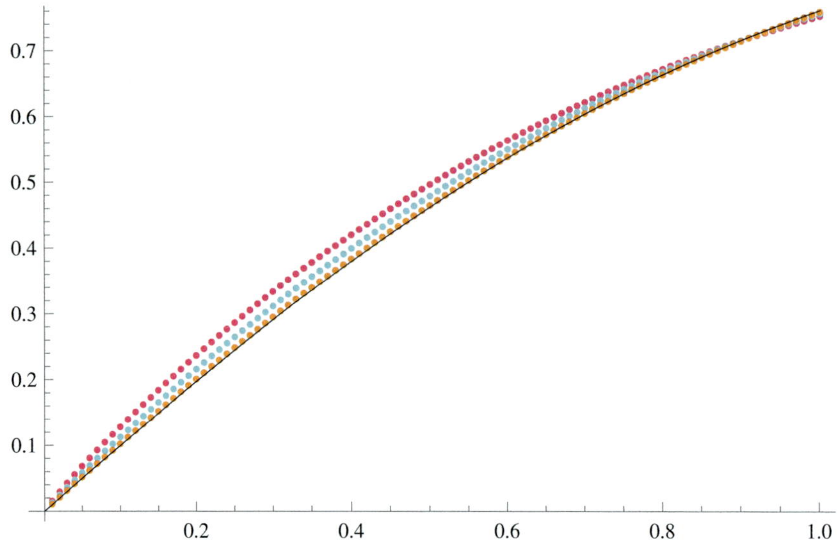

Figure 5.5 Comparison of the exact solution (black line) and the approximate solutions $y(t)$ for $h = 0.01$ when $\alpha = 0.99$ (....), $\alpha = 0.95$ (....) and $\alpha = 0.85$ (....) for Example 5.4.

TABLE 5.3 Absolute Errors of $y(t)$ for $\alpha = 0.99$ and the Different Values of h for Example 5.4

t	$h = 0.1$	$h = 0.01$
0	0	0
0.1	1.04855×10^{-3}	2.46598×10^{-3}
0.2	4.09083×10^{-4}	3.4659×10^{-3}
0.3	1.41762×10^{-3}	3.70756×10^{-3}
0.4	4.01614×10^{-3}	3.40727×10^{-3}
0.5	6.99742×10^{-3}	2.72275×10^{-3}
0.6	1.00206×10^{-2}	1.79206×10^{-3}
0.7	1.28168×10^{-2}	7.35785×10^{-4}
0.8	1.52011×10^{-2}	3.46845×10^{-4}
0.9	1.70695×10^{-2}	1.38051×10^{-3}
1	1.83871×10^{-2}	2.31366×10^{-3}

TABLE 5.4 Absolute Errors of $y(t)$ for $h = 0.01$ and the Different Values of α for Example 5.4

t	$\alpha = 0.95$	$\alpha = 0.99$
0	0	0
0.1	1.32003×10^{-2}	2.46598×10^{-3}
0.2	1.87079×10^{-2}	3.4659×10^{-3}
0.3	2.05325×10^{-2}	3.70756×10^{-3}
0.4	1.97763×10^{-2}	3.40727×10^{-3}
0.5	1.71832×10^{-2}	2.72275×10^{-3}
0.6	1.33766×10^{-2}	1.79206×10^{-3}
0.7	8.89037×10^{-3}	7.35785×10^{-4}
0.8	4.16261×10^{-3}	3.46845×10^{-4}
0.9	4.70838×10^{-4}	1.38051×10^{-3}
1	4.77346×10^{-3}	2.31366×10^{-3}

5.3 GRÜNWALD-LETNIKOV'S APPROACH

In this section, we apply the Grünwald-Letnikov's approach to solve fractional differential equation.

We recall that the Grünwald-Letnikov fractional derivative is defined as Ref. [92]

$$ {}^{G}D^{\alpha}y(t) = \lim_{h \longrightarrow 0} \frac{1}{h^{\alpha}} \sum_{i=0}^{[\frac{t}{h}]} (-1)^{i} \binom{\alpha}{i} y(t - ih), \tag{5.27} $$

and the shifted Grünwald-Letnikov fractional derivative estimate is defined as

$$ {}^{G}D^{\alpha}y(t) = \lim_{h \longrightarrow 0} \frac{1}{h^{\alpha}} \sum_{i=0}^{[\frac{t}{h}]+1} (-1)^{i} \binom{\alpha}{i} y(t - (i-1)h), \tag{5.28} $$

where [.] means the integer part.

Lemma 5.1 *Assume that y satisfies some smoothness conditions e.g., $y(t)$ can be written in the form of a power series for $|t| < \rho$. The Grünwald-Letnikov formula holds for each $0 < \tau < \rho$ and a series of step sizes h with $\frac{\tau}{h} \in \mathbb{N}$ [92]*

$$ \mathbb{D}^{\alpha}y(\tau) = \frac{1}{h^{\alpha}} \Delta_{h}^{\alpha} y(nh) + O(h), \qquad (h \to 0), $$

where

$$ \Delta_{h}^{\alpha} y(nh) = \sum_{i=0}^{n} (-1)^{i} \binom{\alpha}{i} y(t_{n-i}). \tag{5.29} $$

In the case of Caputo's operator, we have according to Theorem 1.1, for $0 < \alpha < 1$,

$$ D^{\alpha}y(\tau) = \frac{1}{h^{\alpha}} \Delta_{h}^{\alpha} y(nh) - \frac{\tau^{-\alpha}}{\Gamma(1-\alpha)} y(0) + O(h), \qquad (h \to 0). \tag{5.30} $$

The most favorable case is when the initial values for Caputo's differential equation are given to be zero.

Here, The Grünwald-Letnikov's approach is applied to approximate the solution of fractional order Van der Pol equation as Ref. [50]

$$D^{\alpha}y(t) + \mu(y^2(t) - 1)y'(t) + y(t) = 0, \quad 1 < \alpha < 2, \tag{5.31}$$

with the state space formulation as

$$\begin{cases} D^{\alpha}y_1(t) = y_2(t), \quad 0 < \alpha < 1, \\ \dfrac{dy_2(t)}{dt} = -y_1(t) - \mu\left(y_1^2(t) - 1\right)y_2(t). \end{cases} \tag{5.32}$$

First, for system (5.32), we consider

$$\begin{cases} D^{\alpha}y_1(t_n) = y_2(t_n), \quad 0 < \alpha < 1, \\ \dfrac{dy_2(t_n)}{dt} = -y_1(t_n) - \mu\left(y_1^2(t_n) - 1\right)y_2(t_n), \end{cases} \tag{5.33}$$

with initial conditions

$$y_i(0) = c_i, \quad i = 1, 2, \quad 0 < \alpha < 1. \tag{5.34}$$

Second, we use a uniform grid $t_n = nh$, where $n = 0, 1, \ldots, N$, $Nh = T$ and we use the abbreviation y_n for approximation of the true solution $y(t_n)$ in the grid point t_n.

We now approximate the first-order derivative $\frac{dy(t)}{dt}$ by using

$$\frac{dy(t)}{dt} = \frac{y(t+h) - y(t)}{h} + O(h), \quad (h \to 0). \tag{5.35}$$

Applying the shifted definition of the Grünwald-Letnikov fractional derivative and Eq. (5.30) to our system (5.32), we obtain

$$\begin{cases} \frac{1}{h^{\alpha}}\displaystyle\sum_{n=0}^{n+1}(-1)^i\binom{\alpha}{i}y_{1,n+1-i} - \dfrac{(nh)^{-\alpha}}{\Gamma(1-\alpha)}y_{10} = y_{2n}, \\ \\ \dfrac{y_{2(n+1)} - y_{2n}}{h} = -y_{1n} - \mu\left(y_{1n}^2 - 1\right)y_{2n}, \end{cases} \tag{5.36}$$

with initial conditions

$$y_i(0) = c_i, \quad i = 1, 2, \quad 0 < \alpha < 1. \tag{5.37}$$

5.3.1 Applications

Example 5.5 *We consider the following fractional Van der Pol equation [50]*

$$\begin{cases} D^{\alpha}y_1(t) = y_2(t), \\ \dfrac{dy_2(t)}{dt} = 2\cos t - \cos^3 t - y_1(t) - y_1^2(t)y_2(t) - y_2(t), \end{cases} \tag{5.38}$$

with the initial conditions

$$y(0) = 0, \quad y^{'}(0) = 1, \tag{5.39}$$

The exact solution, when $\alpha = 1$, *is*

$$y_1(t) = \sin t, \quad y_2(t) = \cos t.$$

According to Eqs. (5.36) and (5.39), we can write Eq. (5.38) as

$$\begin{cases} \dfrac{1}{h^{\alpha}} \displaystyle\sum_{i=0}^{n+1} (-1)^i \binom{\alpha}{i} y_1(t_{n+1-i}) = y_2(t_n), \\ \\ \dfrac{y_2(t_{n+1}) - y_2(t_n)}{h} = 2\cos(t_n) - \cos^3(t_n) - y_1(t_n) - y_1^2(t_n) y_2(t_n) - y_2(t_n), \end{cases} \tag{5.40}$$

We have taken the step size $h = 0.1$, 0.01and $h = 0.001$ in this example. The results of our computation agree with the analytical solution. Figures 5.6 and 5.7 show exact solution and the solution obtained $y_1(t)$ and $y_2(t)$ of Eq. (5.10) for $h = 0.01$ and the different values of α, respectively. Tables 5.5 and 5.6 show the absolute errors of $y_1(t)$ and $y_2(t)$ for $\alpha = 1$ and the different values of h. It is obvious that the numerical solution to converge to the exact solution when the step size h becomes small.

Example 5.6 *Consider the following fractional differential equation*

$$D^2 y(t) + t^{3.5}\, D^{1.5} y(t) + y^2(t) = \left(1 + \frac{\sqrt{\pi}}{4}\right) t^4 + 2, \tag{5.41}$$

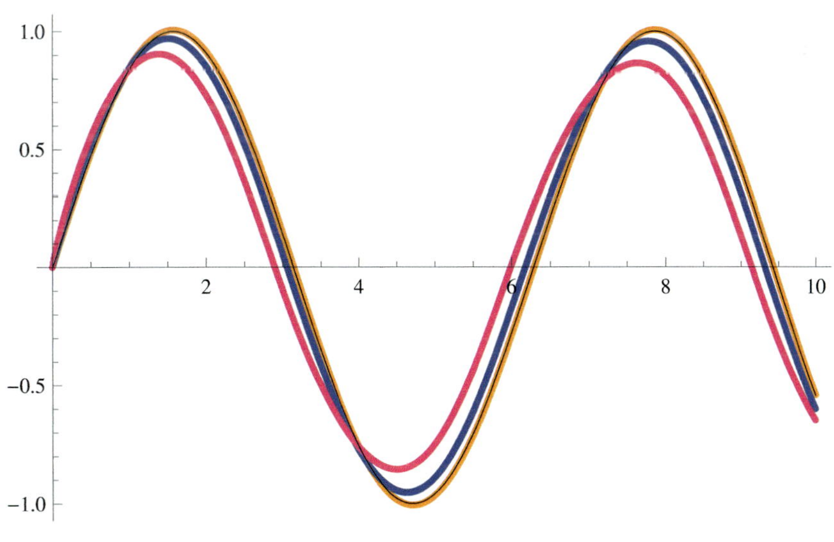

Figure 5.6 Comparison of the exact solution (black line) and the approximate solutions $y_1(t)$ for $h = 0.01$ when $\alpha = 1$ (yellow line), $\alpha = 0.95$ (blue line), $\alpha = 0.85$ (red line) for Example 5.5.

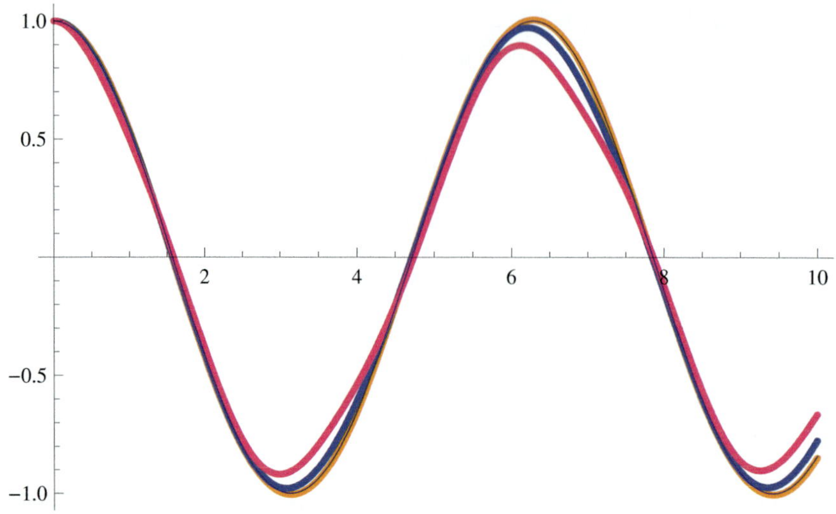

Figure 5.7 Comparison of the exact solution (black line) and the approximate solutions $y_2(t)$ for $h = 0.01$ when $\alpha = 1$ (yellow line), $\alpha = 0.95$ (blue line), $\alpha = 0.85$ (red line) for Example 5.5.

TABLE 5.5 Absolute Errors of $y_1(t)$ for the Different Values of h for Example 5.5

t	$h = 0.5$	$h = 0.1$	$h = 0.01$
0.5	2.05745×10^{-2}	1.00637×10^{-2}	1.0859×10^{-3}
1.0	1.58529×10^{-1}	3.42687×10^{-2}	3.36921×10^{-3}
1.5	3.34828×10^{-1}	5.55561×10^{-2}	5.38257×10^{-3}
2.0	4.03745×10^{-1}	6.97189×10^{-2}	6.83362×10^{-3}
2.5	4.24242×10^{-1}	7.87818×10^{-2}	7.71671×10^{-3}
3.0	4.68387×10^{-1}	7.65612×10^{-2}	7.23235×10^{-3}
3.5	4.42082×10^{-1}	4.98823×10^{-2}	4.1898×10^{-3}
4.0	2.8045×10^{-1}	1.44461×10^{-3}	7.40791×10^{-4}
4.5	6.60281×10^{-2}	4.86126×10^{-2}	4.84581×10^{-3}
5.0	4.41801×10^{-1}	7.4268×10^{-2}	7.21529×10^{-3}
5.5	5.21455×10^{-1}	8.7575×10^{-2}	8.54876×10^{-3}
6.0	5.4484×10^{-1}	9.08809×10^{-2}	8.68595×10^{-3}
6.5	5.69064×10^{-1}	7.2345×10^{-2}	6.38024×10^{-3}
7.0	4.70691×10^{-1}	2.34396×10^{-2}	1.43607×10^{-3}
7.5	1.71494×10^{-1}	3.29192×10^{-2}	3.51687×10^{-3}
8.0	2.80439×10^{-1}	6.75997×10^{-2}	6.56766×10^{-3}
8.5	5.46384×10^{-1}	8.47682×10^{-2}	8.25761×10^{-3}
9.0	5.20337×10^{-1}	9.21087×10^{-2}	8.87914×10^{-3}
9.5	5.91684×10^{-1}	8.15529×10^{-2}	7.41522×10^{-3}
10	5.33826×10^{-1}	4.05461×10^{-2}	3.05155×10^{-3}

TABLE 5.6 Absolute Errors of $y_2(t)$ for the Different Values of h for Example 5.5

t	$h = 0.5$	$h = 0.1$	$h = 0.01$
0.5	1.22417×10^{-1}	1.74806×10^{-2}	1.56591×10^{-3}
1.0	1.24345×10^{-1}	5.8635×10^{-3}	3.8534×10^{-4}
1.5	1.09299×10^{-1}	1.70509×10^{-2}	1.59859×10^{-3}
2.0	1.6451×10^{-1}	2.20787×10^{-2}	2.11047×10^{-3}
2.5	2.52694×10^{-2}	2.01886×10^{-2}	2.18484×10^{-3}
3.0	4.64244×10^{-2}	3.52294×10^{-2}	3.99155×10^{-3}
3.5	1.98845×10^{-1}	6.96556×10^{-2}	7.01571×10^{-3}
4.0	4.80769×10^{-1}	6.97345×10^{-2}	6.04539×10^{-3}
4.5	5.03538×10^{-1}	1.79953×10^{-2}	1.36815×10^{-3}
5.0	6.37972×10^{-2}	1.41194×10^{-2}	1.29036×10^{-3}
5.5	9.68093×10^{-2}	1.72435×10^{-2}	1.77203×10^{-3}
6.0	1.95463×10^{-2}	2.68218×10^{-2}	3.13667×10^{-3}
6.5	1.0389×10^{-1}	6.28786×10^{-2}	6.66618×10^{-3}
7.0	4.0652×10^{-1}	8.55218×10^{-2}	7.71648×10^{-3}
7.5	6.59946×10^{-1}	4.22224×10^{-2}	3.34961×10^{-3}
8.0	2.95648×10^{-1}	5.32397×10^{-3}	5.36534×10^{-4}
8.5	2.22821×10^{-1}	1.59348×10^{-2}	1.55084×10^{-3}
9.0	7.92853×10^{-2}	2.10238×10^{-2}	2.43517×10^{-3}
9.5	5.62838×10^{-2}	5.06219×10^{-2}	5.58536×10^{-3}
10	3.06791×10^{-1}	8.56211×10^{-2}	8.07842×10^{-3}

with initial conditions

$$y(0) = 0, \quad y'(0) = 0, \tag{5.42}$$

the exact solution is

$$y(t) = t^2,$$

We write Eq. (5.41) as the system

$$\begin{cases} y(t) = y_1(t), \\ D^{0.5}y_1(t) = y_2(t), \\ D^{0.5}y_2(t) = y_3(t), \\ D^{0.5}y_3(t) = y_4(t), \\ D^{0.5}y_4(t) = \left(1 + \frac{\sqrt{\pi}}{4}\right) t^4 - y_1^2(t) - t^{3.5}y_3(t) + 2. \end{cases} \tag{5.43}$$

We have taken the step size $h = 0.1$, 0.01 and $h = 0.001$ in this example. The results of our computation agree with the analytical solution.

Figures 5.8 and 5.9 show exact solution and the solution obtained $y(t)$ of Eq. (5.41) for $h = 0.01$ and $h = 0.001$, respectively. Table 5.7 shows the absolute errors of $y(t)$ for the different values of h. It is obvious that the numerical solution to converge to the exact solution when the step size h becomes small.

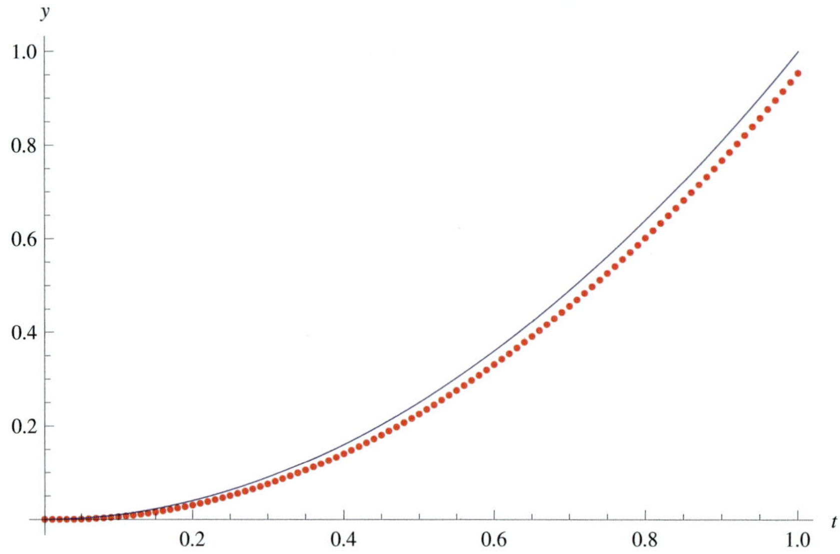

Figure 5.8 Comparison of the exact solution (black line) and the approximate solutions $y(t)$ when $h = 0.01$ (....), for Example 5.6

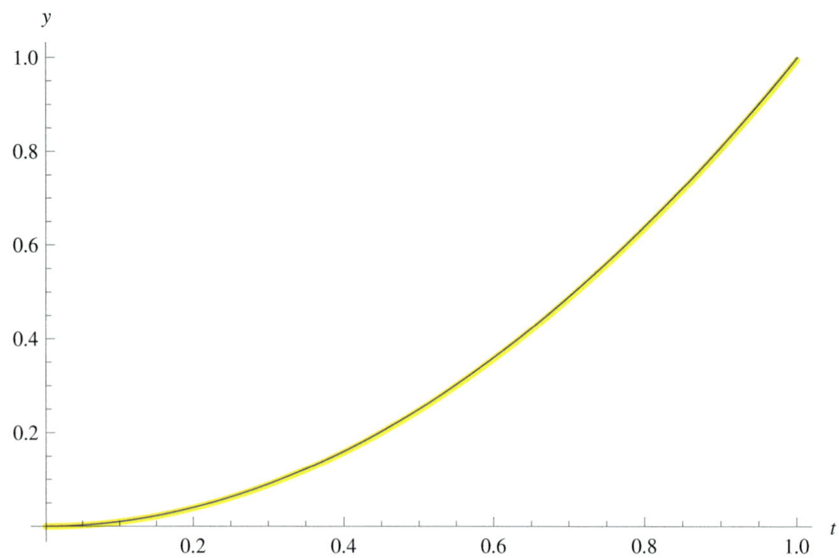

Figure 5.9 Comparison of the exact solution (black line) and the approximate solutions $y(t)$ when $h = 0.001$ (....), for Example 5.6.

Also, there are other numerical methods such as redial basis function, finite difference, finite volume methods and so on for solving fractional differential equations [25, 30, 62, 109, 115].

TABLE 5.7 Absolute Errors of $y(t)$ for the Different Values of h for Example 5.6

t	$h = 0.1$	$h = 0.01$	$h = 0.001$
0.1	1.0000×10^{-1}	4.4000×10^{-3}	4.93996×10^{-4}
0.2	4.0000×10^{-1}	9.39949×10^{-3}	9.93852×10^{-4}
0.3	9.0000×10^{-1}	1.43941×10^{-2}	1.49284×10^{-3}
0.4	1.4000×10^{-1}	1.93698×10^{-2}	1.98903×10^{-3}
0.5	1.89999×10^{-1}	2.4297×10^{-2}	2.47875×10^{-3}
0.6	2.39979×10^{-1}	2.9126×10^{-2}	2.95621×10^{-3}
0.7	2.89872×10^{-1}	3.37846×10^{-2}	3.41337×10^{-3}
0.8	3.39495×10^{-1}	3.81801×10^{-2}	3.8404×10^{-3}
0.9	3.88504×10^{-1}	4.22073×10^{-2}	4.22668×10^{-3}
1	4.36343×10^{-1}	4.57629×10^{-2}	4.5625×10^{-3}

CHAPTER 6

Operational Matrices

RECENTLY, operational matrices were applied for solving several kinds of fractional differential equations (FDEs). The use of numerical techniques in conjunction with operational matrices of orthogonal or non-orthogonal polynomials for the solution of FDEs on finite and infinite intervals has produced highly accurate solutions for such equations. Orthogonal or non-orthogonal polynomials play a crucial role in spectral methods. In this chapter, we consider particular cases of these polynomials such as the Jacobi polynomials, Chebyshev polynomials, Bernstein polynomials and Genocchi polynomials. This chapter discusses spectral methods based on operational matrices of fractional derivatives and integrals for solving several kinds of linear and nonlinear fractional differential equations.

We present the operational matrices of fractional derivatives and integrals, for several polynomials such as the Jacobi, Chebyshev, Bernstein and Genocchi polynomials, and we use them with different spectral techniques for solving the aforementioned equations.

In recent years, there has been considerable interest in employing operational matrix for numerically solving many types of integral and differential equations [8, 31, 60, 65, 69, 84, 97, 100, 101, 103, 117]. The operational matrix converts the considered equation in a system of linear algebraic equations. The speed of convergence is one of the great advantages of spectral methods. Besides, spectral methods have exponential rates of convergence and high level of accuracy. Spectral methods can be classified into three types, namely the collocation, tau and Galerkin methods. As it is well known, one of the most accurate methods of discretization for solving numerous differential equations is spectral method. The spectral methods based on polynomials, such as Bernstein polynomials, Jacobi polynomials and their special cases, are only available for bounded domains for approximation of FDEs [10–12, 65, 66, 102]. The use of spectral methods based on orthogonal polynomials, such as the modified generalized Laguerre polynomials and their special cases, is available for unbounded domains for approximation of FDEs [99].

A sequence of polynomials $\{p_n(x)\}_0^\infty$, each with degree n is called orthogonal with respect to the weight function $w(x)$ on the interval $[a, b] \in \mathbb{R}$ with $a < b$ if

$$\int_a^b w(x)p_n(x)p_m(x)dx = \delta_{nm}c_n,$$

DOI: 10.1201/9781003474609-6

where

$$\delta_{nm} = \begin{cases} 0, & n \neq m, \\ 1, & n = m. \end{cases}$$

The weight function $w(x)$ should be continuous and positive on $[a, b]$. If $c_n = 1$, the polynomials are called orthonormal.

6.1 JACOBI POLYNOMIALS

The Jacobi polynomials have become increasingly important in numerical analysis, from both theoretical and practical points of view. One of the common sets of orthogonal polynomials is known as Jacobi polynomials which are defined by means of Rodrigues formula [15, 36]

$$P_n^{(\nu,\gamma)}(z) = \frac{(-1)^n}{2^n n!}(1-z)^{-\nu}(1+z)^{-\gamma}\frac{d^n}{dz^n}(1-z)^{n+\nu}(1+z)^{n+\gamma},$$

The Jacobi polynomials satisfy the following Sturm-Liouville problem

$$(1-z)^{-\nu}(1+z)^{-\gamma}\frac{d}{dz}\left\{(1-z)^{\nu+1}(1+z)^{\gamma+1}\frac{d}{dz}P_n^{(\nu,\gamma)}(z)\right\}$$
$$+n(n+1+\nu+\gamma)P_n^{(\nu,\gamma)}(z) = 0.$$

The three-term recurrence relation for the Jacobi polynomials is given by

$$\begin{aligned} P_0^{(\nu,\gamma)}(z) &= 1, \\ P_1^{(\nu,\gamma)}(z) &= A_0 z + B_0, \\ P_{n+1}^{(\nu,\gamma)}(z) &= (A_n z + B_n)P_n^{(\nu,\gamma)}(z) - C_n P_{n-1}^{(\nu,\gamma)}(z), \qquad n \geq 1, \end{aligned}$$

where the constant coefficients satisfy

$$\begin{aligned} A_0 &= \frac{1}{2}(\nu+\gamma)+1, \\ B_0 &= \frac{1}{2}(\nu-\gamma), \\ A_n &= \frac{(2n+\nu+\gamma+1)(2n+\nu+\gamma+2)}{2(n+1)(n+\nu+\gamma+1)}, \\ B_n &= \frac{(\nu^2-\gamma^2)(2n+\nu+\gamma+1)}{2(n+1)(n+\nu+\gamma+1)(2n+\nu+\gamma)}, \\ C_n &= \frac{(n+\nu)(n+\gamma)(2n+\nu+\gamma+2)}{(n+1)(n+\nu+\gamma+1)(2n+\nu+\gamma)}. \end{aligned}$$

These polynomials are orthogonal with respect to the weight function $w(x) = (1-z)^{\nu}\ (1+z)^{\gamma}$, $\nu, \gamma > -1$ on the interval $[-1, 1]$ with respect to the inner product

$$\left\langle P_n^{(\nu,\gamma)}, P_m^{(\nu,\gamma)} \right\rangle = \int_{-1}^{1} w(z)P_n^{(\nu,\gamma)}(z)P_m^{(\nu,\gamma)}(z)dz.$$

In fact, we have

$$\left\langle P_n^{(\nu,\gamma)}, P_m^{(\nu,\gamma)} \right\rangle = \delta_{nm} h_n,$$

for $m, n \in \mathbb{N}_0 = \mathbb{N} \cup \{0\}$ where

$$h_n = \frac{2^{\nu+\gamma+1}\Gamma(\nu+n+1)\Gamma(\gamma+n+1)}{n!(\nu+\gamma+2n+1)\Gamma(\nu+\gamma+n+1)}.$$

The analytic form of Jacobi polynomials is given by

$$P_n^{(\nu,\gamma)}(z) = \sum_{k=0}^{n} \frac{(-1)^{n-k}\Gamma(n+\gamma+1)\Gamma(n+k+\nu+\gamma+1)}{\Gamma(k+\gamma+1)\Gamma(n+\nu+\gamma+1)(n-k)!k!} \left(\frac{1+z}{2}\right)^k, \quad n \geq 1.$$

For practical use of Jacobi polynomials on the interval $t \in [0,1]$, it is necessary to shift the defining domain by means of the following change variable:

$$z = 2t - 1, \quad t \in [0,1].$$

The shifted Jacobi polynomials can be generated from the following recurrence relation

$$\begin{aligned}
&P_0^{(\nu,\gamma)}(t) = 1, \\
&P_1^{(\nu,\gamma)}(t) = A_0(2t-1) + B_0, \\
&P_{n+1}^{(\nu,\gamma)}(t) = (A_n(2t-1) + B_n)P_n^{(\nu,\gamma)}(t) - C_n P_{n-1}^{(\nu,\gamma)}(t), \qquad n \geq 1. \quad (6.1)
\end{aligned}$$

The orthogonality condition is given by

$$\int_0^1 w(t)P_n^{(\nu,\gamma)}(t)P_m^{(\nu,\gamma)}(t)dt = \delta_{nm}\theta_n, \tag{6.2}$$

where

$$\theta_n = \frac{\Gamma(n+\nu+1)\Gamma(n+\gamma+1)}{n!(2n+\nu+\gamma+1)\Gamma(n+\nu+\gamma+1)},$$

and

$$w(t) = (1-t)^{\nu} t^{\gamma}, \quad t \in [0,1].$$

Also, the analytic form of shifted Jacobi polynomials will be as follows:

$$P_n^{(\nu,\gamma)}(t) = \sum_{k=0}^{n} \frac{(-1)^{n-k}\Gamma(n+\gamma+1)\Gamma(n+k+\nu+\gamma+1)t^k}{\Gamma(k+\gamma+1)\Gamma(n+\nu+\gamma+1)(n-k)!k!}, \quad n \geq 1. \tag{6.3}$$

Some other properties of the shifted Jacobi polynomials are presented as follows:

$$P_n^{(\nu,\gamma)}(0) = (-1)^n \frac{\Gamma(n+\gamma+1)}{\Gamma(\gamma+1)n!},$$

$$\frac{d^i}{dt^i}P_n^{(\nu,\gamma)}(0)=\frac{(-1)^{n-i}\Gamma(n+\gamma+1)(n+\nu+\gamma+1)_i}{\Gamma(n-i+1)\Gamma(i+\gamma+1)}, \tag{6.4}$$

A function $y(t)$, square integrable in $[0,1)$, can be expressed in terms of shifted Jacobi polynomials as

$$y(t)=\sum_{j=0}^{\infty}c_jP_j^{(\nu,\gamma)}(t),$$

where the coefficients c_j are given by

$$c_j=\frac{1}{\theta_j}\int_0^1 y(t)P_j^{(\nu,\gamma)}(t)w(t)dt, \quad j=0,1,\ldots. \tag{6.5}$$

In practice, only the first $(n+1)$-terms of shifted Jacobi polynomials are considered. Therefore, one has

$$y_n(t)\simeq\sum_{j=0}^{n}c_jP_j^{(\nu,\gamma)}(t)=C^T\Phi(t), \tag{6.6}$$

where the vectors C and $\Phi(x)$ are given by

$$C=[c_0,c_1,\ldots,c_n]^T, \qquad \Phi(t)=\left[P_0^{(\nu,\gamma)}(t),P_1^{(\nu,\gamma)}(t),\ldots,P_n^{(\nu,\gamma)}(t)\right]^T. \tag{6.7}$$

6.1.1 Operational Matrix of Fractional Integration

In this section, operational matrices of the shifted Jacobi will be derived. To this end, some useful lemmas and theorems are stated.

Let $\Phi(x)$ be the shifted Jacobi vector and $\alpha>0$ then [11, 12, 68]

$$I^{\alpha}\Phi(t)\simeq P^{\alpha}\Phi(t), \tag{6.8}$$

where P^{α} is the $(n+1)\times(n+1)$ operational matrix of fractional integration of order α in the Riemann-Liouville sense.

By applying the Riemann-Liouville integral operator to the analytic form of Jacobi polynomials (6.3), we calculate P^{α} as follows:

$$\begin{aligned} I^{\alpha}P_i^{(\nu,\gamma)}(t) &= \sum_{k=0}^{i}\frac{(-1)^{i-k}\Gamma(i+\gamma+1)\Gamma(i+k+\nu+\gamma+1)}{\Gamma(k+\gamma+1)\Gamma(i+\nu+\gamma+1)(i-k)!k!}I^{\alpha}t^k \\ &= \sum_{k=0}^{i}\frac{(-1)^{i-k}\Gamma(i+\gamma+1)\Gamma(i+k+\nu+\gamma+1)}{\Gamma(k+\gamma+1)\Gamma(i+\nu+\gamma+1)(i-k)!\Gamma(k+\alpha+1)}t^{k+\alpha}, \\ & i=0,1,\ldots,n. \end{aligned} \tag{6.9}$$

Now, approximate $t^{k+\alpha}$ by $n+1$ terms of shifted Jacobi series, yields

$$t^{k+\alpha}=\sum_{j=0}^{n}c_{kj}P_i^{(\nu,\gamma)}(t), \tag{6.10}$$

where c_{kj} is obtained as

$$
\begin{aligned}
c_{kj} &= \frac{(2j+\nu+\gamma+1)j!}{\Gamma(j+\nu+1)} \\
&\times \sum_{r=0}^{j} \frac{(-1)^{j-r}\Gamma(j+r+\nu+\gamma+1)\Gamma(\nu+1)\Gamma(r+k+\gamma+\alpha+1)}{\Gamma(r+\gamma+1)\Gamma(r+k+\nu+\gamma+\alpha+2)(j-r)!r!}, \\
& j = 1, 2, \ldots, n.
\end{aligned}
$$

In virtue of Eqs. (6.9) and (6.10), we get

$$
I^{\alpha} P_i^{(\nu,\gamma)}(t) = \sum_{j=0}^{n} \Omega_{\alpha}(i,j) P_i^{(\nu,\gamma)}(t), \quad i = 0, 1, \ldots, n. \tag{6.11}
$$

where $\Omega_{\alpha}(i,j) = \sum_{k=0}^{i} \zeta_{ijk}$, and

$$
\begin{aligned}
\zeta_{ijk} &= \frac{(-1)^{i-k}\Gamma(i+\gamma+1)\Gamma(i+k+\nu+\gamma+1)}{\Gamma(k+\gamma+1)\Gamma(i+\nu+\gamma+1)(i-k)!\Gamma(k+\nu+1)} \\
&\times \sum_{r=0}^{j} \frac{(-1)^{j-r}\Gamma(j+r+\nu+\gamma+1)\Gamma(\nu+1)\Gamma(r+k+\gamma+\alpha+1)(2j+\nu+\gamma+1)j!}{\Gamma(j+\nu+1)\Gamma(r+\gamma+1)\Gamma(r+k+\nu+\gamma+\alpha+2)(j-r)!r!}, \\
& j = 1, 2, \ldots, n.
\end{aligned}
$$

Accordingly, Eq. (6.11) can be written in a vector form as follows:

$$
I^{\alpha} P_i^{(\nu,\gamma)}(t) \simeq [\Omega_{\alpha}(i,0), \Omega_{\alpha}(i,1), \ldots, \Omega_{\alpha}(i,n)]\, \Phi(t), \quad i = 0, 1, \ldots, n,
$$

and the operational matrix of fractional integration of order α is as

$$
P^{\alpha} = \begin{bmatrix}
\Omega_{\alpha}(0,0) & \Omega_{\alpha}(0,1) & \ldots & \Omega_{\alpha}(0,n) \\
\Omega_{\alpha}(1,0) & \Omega_{\alpha}(1,1) & \ldots & \Omega_{\alpha}(1,n) \\
\vdots & \vdots & \ldots & \vdots \\
\Omega_{\alpha}(i,0) & \Omega_{\alpha}(i,1) & \ldots & \Omega_{\alpha}(i,n) \\
\vdots & \vdots & \ldots & \vdots \\
\Omega_{\alpha}(n,0) & \Omega_{\alpha}(n,1) & \ldots & \Omega_{\alpha}(n,n)
\end{bmatrix}. \tag{6.12}
$$

6.1.2 Applications

This section is dedicated to introducing a new numerical method for solving the multi-term FDE [12]

$$
D^{\alpha} y(t) = \sum_{i=1}^{k} \mu_i D^{\beta_i} y(t) + \mu_{k+1} y(t) + f(t), \tag{6.13}
$$

with initial conditions

$$
y^{(i)}(0) = d_i, \quad i = 0, \ldots, m-1, \tag{6.14}
$$

where $\mu_i (i = 1, 2, \ldots, k)$ are real constant coefficients and also $m - 1 < \alpha < m$, $0 < \beta_1 < \beta_2 < \ldots < \beta_k < \alpha$. The values of $d_i (i = 0, \ldots, m-1)$ describe the initial state of $y(t)$ and $f(t)$ is a given source function.

By applying the Riemann-Liouville integral of order α on Eq. (6.13) and after making use of initial conditions, we get the following form

$$y(t) - \sum_{j=0}^{m-1} y^{(j)}(0)\frac{t^j}{j!} = \sum_{i=1}^{k} \mu_i I^{\alpha-\beta_i}\left(y(t) - \sum_{j=0}^{m_i-1} y^{(j)}(0)\frac{t^j}{j!}\right) + \mu_{k+1} I^{\alpha} y(t) + I^{\alpha} f(t), \tag{6.15}$$

where $m_i - 1 < \beta_i \leq m_i$, $m_i \in \mathbb{N}$, this implies that

$$y(t) = \sum_{i=1}^{k} \mu_i I^{\alpha-\beta_i} y(t) + \mu_{k+1} I^{\alpha} y(t) + g(t), \tag{6.16}$$

where

$$g(t) = I^{\alpha} f(t) + \sum_{j=0}^{m-1} d_j \frac{t^j}{j!} + \sum_{i=1}^{k} \mu_i I^{\alpha-\beta_i}\left(\sum_{j=0}^{m_i-1} d_j \frac{t^j}{j!}\right). \tag{6.17}$$

Now, we approximate $y(t)$ and $g(t)$ by means of the shifted Jacobi polynomials:

$$y(t) \simeq \sum_{i=0}^{n} c_i P_i^{(\nu,\gamma)}(t) = C^T \Phi(t), \tag{6.18}$$

$$g(t) \simeq \sum_{i=0}^{n} g_i P_i^{(\nu,\gamma)}(t) = G^T \Phi(t), \tag{6.19}$$

where

$$C = [c_0, c_1, \ldots, c_n]^T$$
$$G = [g_0, g_1, \ldots, g_n]^T$$

Also, we approximate the Riemann-Liouville integral of orders α and $\alpha - \beta_i$ in Eq. (6.16)

$$I^{\alpha} y(t) \simeq C^T I^{\alpha} \Phi(t) \simeq C^T P^{\alpha} \Phi(t), \tag{6.20}$$

$$I^{\alpha-\beta_i} y(t) \simeq C^T I^{\alpha-\beta_i} \Phi(t) \simeq C^T P^{\alpha-\beta_i} \Phi(t), \quad i = 1, \ldots, k, \tag{6.21}$$

where P^{α} is the $(n+1) \times (n+1)$ operational matrix of fractional integration of order α.

Employing equations (6.18)–(6.21), the residual $R_n(t)$ for Eq. (6.16) can be written as

$$R_n(t) = \left(C^T - C^T \sum_{j=1}^{k} \mu_j P^{\alpha-\beta_j} - \mu_{k+1} C^T P^{\alpha} - G^T\right) \Phi(t).$$

As in a typical tau method, we generate $n-m+1$ linear algebraic equations by applying

$$< R_n(t), P_j^{(\nu,\gamma)}(t) >= \int_0^1 R_n(t)P_j^{(\nu,\gamma)}(t)dt = 0, \quad j = 0, 1, \dots, n-m.$$

Also by substituting equations (6.4) and (6.18) in Eq. (6.14), we get

$$y^{(i)}(0) = \sum_{j=0}^{n} c_j D^{(i)} P_j^{(\nu,\gamma)}(0) = d_i, \quad i = 0, 1, \dots, m-1. \tag{6.22}$$

Equations (6.22) and (6.22) generate $n-m+1$ and m set of linear equations, respectively. These linear equations can be solved for unknown coefficients of the vector C. Consequently, $y(t)$ given in Eq. (6.18) can be calculated, which gives a solution of equation (6.13) with the initial conditions (6.14).

Example 6.1 *we consider the following initial value problem [12]*

$$D^{\frac{3}{2}}y(t) + 3y(t) = 3t^3 + \frac{8}{\Gamma(0.5)}t^{1.5}, \tag{6.23}$$

$$y(0) = y'(0) = 0, \tag{6.24}$$

whose exact solution is given by $y(t) = t^3$.

We apply the Riemann-Liouville integral of order $\frac{3}{2}$ on Eq. (6.23), we have

$$y(t) + 3I^{\frac{3}{2}}y(t) = \frac{64t^{\frac{9}{2}}}{105\sqrt{\pi}} + t^3. \tag{6.25}$$

By applying the technique described in Section 6.1.2 with $n = 3$, we may write the approximate solution and the right-hand side in the forms

$$y(t) \simeq \sum_{i=0}^{3} c_i p^{(\nu,\gamma)}(t) = C^T\Phi(t), \qquad g(t) \simeq \sum_{i=0}^{3} g_i p^{(\nu,\gamma)}(t) = G^T\Phi(t), \tag{6.26}$$

where $g(t) = \frac{64t^{\frac{9}{2}}}{105\sqrt{\pi}} + t^3$. According to Eq. (6.12), the operational matrix of fractional integration of order $\frac{3}{2}$ is as follows:

$$P^{\frac{3}{2}} = \begin{bmatrix} \Omega_{\frac{3}{2}}(0,0) & \Omega_{\frac{3}{2}}(0,1) & \dots & \Omega_{\frac{3}{2}}(0,3) \\ \Omega_{\frac{3}{2}}(1,0) & \Omega_{\frac{3}{2}}(1,1) & \dots & \Omega_{\frac{3}{2}}(1,3) \\ \Omega_{\frac{3}{2}}(2,0) & \Omega_{\frac{3}{2}}(2,1) & \dots & \Omega_{\frac{3}{2}}(2,3) \\ \Omega_{\frac{3}{2}}(3,0) & \Omega_{\frac{3}{2}}(3,1) & \dots & \Omega_{\frac{3}{2}}(3,3) \end{bmatrix},$$

and $G = [g_0, g_1, g_2, g_3]^T$ is as

$$g_j = \frac{(2j+\nu+\gamma+1)j!}{\Gamma(j+\nu+1)} \sum_{r=0}^{j} \frac{(-1)^{j-r}\Gamma(r+j+\nu+\gamma+1)}{r!(j-r)!\Gamma(r+\gamma+1)} \times \int_0^1 (\frac{64t^{\frac{9}{2}}}{105\sqrt{\pi}} + t^3)t^{\gamma+r}(1-t)^{\nu}dt.$$

Employing Eq. (6.26), the residual $R_n(x)$ for Eq. (6.25) can be written as

$$R_n(t) = \left(C^T + 3C^T P^{\frac{3}{2}} - G^T\right)\Phi(t). \tag{6.27}$$

Making use of tau method, we get

$$3\Omega_{\frac{3}{2}}(0,2)c_0 + 3\Omega_{\frac{3}{2}}(1,2)c_1 + 3\Omega_{\frac{3}{2}}(2,2)c_2 + 3\Omega_{\frac{3}{2}}(3,2)c_3 + c_2 - g_2 = 0, \tag{6.28}$$
$$3\Omega_{\frac{3}{2}}(0,3)c_0 + 3\Omega_{\frac{3}{2}}(1,3)c_1 + 3\Omega_{\frac{3}{2}}(2,3)c_2 + 3\Omega_{\frac{3}{2}}(3,3)c_3 + c_3 - g_3 = 0. \tag{6.29}$$

Using the initial conditions (6.24) gives

$$\begin{aligned} C^T\Phi(0) &= c_0 - (\gamma+1)c_1 + \frac{(\gamma+1)(\gamma+2)}{2}c_2 \\ &\quad - \frac{(\gamma+1)(\gamma+2)+(\gamma+3)}{6}c_3 = 0, \end{aligned} \tag{6.30}$$
$$\begin{aligned} C^T D^{(1)}\Phi(0) &= (\nu+\gamma+2)c_1 - (\gamma+2)(\nu+\gamma+3)c_2 \\ &\quad + \frac{(\gamma+2)(\gamma+3)(\nu+\gamma+4)}{2}c_3 = 0. \end{aligned} \tag{6.31}$$

Finally by solving Eqs. (6.28)–(6.31) and the different values of ν and γ, we get the approximate solution.

For $\nu = \gamma = 0$, then

$$c_0 = \frac{1}{4}, \quad c_1 = \frac{9}{20}, \quad c_2 = \frac{1}{4}, \quad c_3 = \frac{1}{20},$$

and the approximate solution is given by

$$y(t) = \sum_{i=0}^{3} c_i P_i^{(0,0)}(t) = t^3,$$

which is the exact solution.

Also if we choose $\nu = -\frac{1}{2}$, $\gamma = \frac{1}{2}$, then

$$c_0 = \frac{35}{64}, \quad c_1 = \frac{21}{32}, \quad c_2 = \frac{7}{24}, \quad c_3 = \frac{1}{20},$$

and

$$y(t) = \sum_{i=0}^{3} c_i P_i^{(-\frac{1}{2},\frac{1}{2})}(t) = t^3,$$

which is the exact solution.

In case of $\nu = \frac{1}{2}$, $\gamma = -\frac{1}{2}$, we have

$$c_0 = \frac{5}{64}, \quad c_1 = \frac{9}{32}, \quad c_2 = \frac{5}{24}, \quad c_3 = \frac{1}{20},$$

and

$$y(t) = \sum_{i=0}^{3} c_i P_i^{(\frac{1}{2},-\frac{1}{2})}(t) = t^3,$$

which is the exact solution.

Example 6.2 *we consider the following initial value problem [12]*

$$\begin{aligned} &D^2y(t) - 2Dy(t) + D^{\frac{1}{2}}y(t) + y(t) = t^7 + \frac{2048}{429\sqrt{\pi}}t^{6.5} - 14t^6 + 42t^5 \\ &\quad -t^2 - \frac{8}{3\sqrt{\pi}}t^{1.5} + 4t - 2, \\ &y(0) = y'(0) = 0, \end{aligned} \tag{6.32}$$

whose the exact solution is given by $y(t) = t^7 - t^2$.

Now, applying the technique described in Example 6.1 with $n = 9$ for some special choices of ν and γ, gives the following cases

If $\nu = \gamma = 0$, then

$$\begin{aligned} &c_0 = \frac{-5}{24}, \quad c_1 = \frac{-5}{24}, \quad c_2 = \frac{3}{24}, \quad c_3 = \frac{49}{264}, \quad c_4 = \frac{7}{88}, \\ &c_5 = \frac{7}{312}, \quad c_6 = \frac{1}{264}, \quad c_7 = \frac{1}{3432}, \quad c_8 = 0, \quad c_9 = 0, \end{aligned}$$

and the exact solution is

$$y(t) = \sum_{i=0}^{9} c_i P^{(0,0)}(t) = t^7 - t^2.$$

If $\nu = \gamma = \frac{1}{2}$, then

$$\begin{aligned} &c_0 = \frac{-1845}{8192}, \quad c_1 = \frac{-1047}{6144}, \quad c_2 = \frac{307}{5120}, \quad c_3 = \frac{13}{128}, \quad c_4 = \frac{25}{567}, \\ &c_5 = \frac{15}{1232}, \quad c_6 = \frac{7}{3432}, \quad c_7 = \frac{1}{6435}, \quad c_8 = 0, \quad c_9 = 0, \end{aligned}$$

and the exact solution is

$$y(t) = \sum_{i=0}^{9} c_i P^{(\frac{1}{2},\frac{1}{2})}(t) = t^7 - t^2.$$

If $\nu = -\frac{1}{2}, \gamma = \frac{1}{2}$, then

$$c_0 = \frac{-3805}{16384}, \quad c_1 = \frac{-115}{8192}, \quad c_2 = \frac{1979}{6144}, \quad c_3 = \frac{273}{1024}, \quad c_4 = \frac{13}{128},$$
$$c_5 = \frac{5}{192}, \quad c_6 = \frac{5}{1232}, c_7 = \frac{1}{3432}, \quad c_8 = 0, \quad c_9 = 0,$$

and the exact solution is

$$y(t) = \sum_{i=0}^{9} c_i P^{(-\frac{1}{2},\frac{1}{2})}(t) = t^7 - t^2.$$

If $\nu = \frac{1}{2}, \gamma = -\frac{1}{2}$, then

$$c_0 = \frac{-1619}{16384}, \quad c_1 = \frac{2071}{8192}, \quad c_2 = \frac{-23}{6144}, \quad c_3 = \frac{637}{5120}, \quad c_4 = \frac{39}{640},$$
$$c_5 = \frac{11}{576}, \quad c_6 = \frac{13}{3696}, \quad c_7 = \frac{1}{3432}, \quad c_8 = 0, \quad c_9 = 0,$$

and the exact solution is

$$y(t) = \sum_{i=0}^{9} c_i P^{(\frac{1}{2},-\frac{1}{2})}(t) = t^7 - t^2.$$

6.2 CHEBYSHEV POLYNOMIALS

In this section, we consider another important special case of the Jacobi polynomials-Chebyshev polynomials (of the first kind), which are proportional to Jacobi polynomials $P^{(-\frac{1}{2},-\frac{1}{2})}$. Chebyshev polynomials belong to the broader family of orthogonal polynomials (ultraspherical polynomials and Jacobi polynomials). In many practical circumstances, the Chebyshev series expansion of a function is superior to all other expansions into ultraspherical polynomials, which have a slower convergence rate compared to Chebyshev polynomials.

The Chebyshev polynomials of the first kind $T_n(t)$ is defined by the relation [79]

$$T_n(t) = \cos n\theta,$$

where $t = \cos\theta$. The elementary recurrence

$$\cos(n+1)(\theta) = 2\cos\theta\cos n\theta - \cos(n-1)\theta,$$

yields the three-term recurrence relation for the Chebyshev polynomials

$$T_{n+1}(t) = 2tT_n(t) - T_{n-1}(t),$$

with $T_0(t) = 1$ and $T_1(t) = t$, shows that $T_n(t)$ is indeed a polynomial in t. The polynomial $T_n(t)$ is of degree n and its leading coefficient is 2^{n-1}.

A useful expression for the Chebyshev polynomials is their Rodrigues formula

$$T_n(t) = \frac{(-1)^n 2^n n!}{(2n)!}\sqrt{1-t^2}\frac{d^n}{dt^n}(1-t^2)^{n-\frac{1}{2}},$$

The Chebyshev polynomials satisfy the following Sturm-Liouville problem

$$\sqrt{1-t^2}\left(\sqrt{1-t^2}T_n'(t)\right)' + n^2 T_n(t) = 0.$$

or equivalently,

$$(1-t^2)T_n''(t) - xT_n'(t) + n^2 T_n(t) = 0.$$

These polynomials are orthogonal with respect to the weight function $w(t) = \frac{1}{\sqrt{1-t^2}}$ on the interval $[-1, 1]$ with respect to the inner product

$$\langle T_n, T_m \rangle = \int_{-1}^{1} w(t)T_n(t)T_m(t)dt = \frac{\pi}{2}c_n\delta_{nm}.$$

where $c_0 = 2$ and $c_1 = 1$ for $n \geq 1$. The analytic form of the Chebyshev polynomials of degree n are given by

$$T_n(t) = \frac{n}{2}\sum_{k=0}^{\lfloor n/2 \rfloor} \frac{(-1)^k \Gamma(n-k)}{\Gamma(k+1)\Gamma(n-2k+1)} t^{n-2k}, \quad n = 1, 2, \ldots,$$

where $\lfloor t \rfloor$ denotes the floor function whose value equals the largest integer less than or equal to t.

The Shifted Chebyshev polynomials of degree n are defined on $t \in [0, 1]$ with the aid of the following recurrence formula: following change variable:

$$T_n^*(t) = 2(2t-1)T_n^*(t) - T_{n-1}^*(t), \quad n = 1, 2, \ldots,$$

where $T_0^*(t) = 1$ and $T_1^*(t) = 2t - 1$. The analytic form of the shifted Chebyshev polynomials $T_n^*(t)$ of degree n is given by

$$T_n^*(t) = n\sum_{k=0}^{n}(-1)^{n-k}\frac{(n+k-1)!2^{2k}}{(n-k)!(2k)!}t^k. \tag{6.33}$$

The orthogonality condition is

$$\int_0^1 w(t)T_n(t)T_m(t)dt = h_n\delta_{nm}.$$

where $w(t) = \frac{1}{\sqrt{t-t^2}}$ and $h_j = \frac{c_j}{2}\pi$, $c_0 = 2$, $c_1 = 1$ for $j \geq 1$.

The special values

$$T_n^{*(q)}(0) = (-1)^{n-q}\frac{n(n+q-1)!}{\Gamma(q+\frac{1}{2})(n-q)!}\sqrt{\pi}, \quad q \leq i, \tag{6.34}$$

will be of important use later.

A function $y(t)$, square integrable in $[0, 1)$, may be expressed in terms of shifted Chebyshev polynomials as

$$y(t) = \sum_{j=0}^{\infty} c_j T_j^*(t),$$

where the coefficients c_j are given by

$$c_j = \frac{1}{h_j} \int_0^1 y(t) T_j^*(t) w(t) dt, \quad j = 0, 1, \ldots. \tag{6.35}$$

In practice, only the first $(n + 1)$-terms of shifted Chebyshev polynomials are considered. Therefore, one has

$$y_n(t) \simeq \sum_{j=0}^{n} c_j T_j^*(t) = C^T \Phi(t), \tag{6.36}$$

where the shifted Chebyshev coefficient vector C and the shifted Chebyshev $\Phi(t)$ are given by

$$C = [c_0, c_1, \ldots, c_n]^T, \quad \Phi(t) = [T_0^*(t), T_1^*(t), \ldots, T_2^*(t)]^T. \tag{6.37}$$

6.2.1 Operational Matrix of Fractional Integration

The main objective of this section is to generalize the operational matrix of fractional integration for the shifted Chebyshev polynomials.

The Riemann-Liouville fractional integral of vector $\Phi(t)$ can be expressed as [10]

$$I^\alpha \Phi(t) \simeq P^\alpha \Phi(t) \tag{6.38}$$

where P^α is the $(n + 1) \times (n + 1)$ operational matrix of fractional integration of order α in the for the Chebyshev polynomials.

In order to obtain the operational matrix of fractional integration, we apply the fractional operator I^α to the function $T_n^*(t)$,

$$\begin{aligned} I^\alpha T_i^*(t) &= i \sum_{k=0}^{i} (-1)^{i-k} \frac{(i + k - 1)!\, 2^{2k}}{(i - k)!(2k)!} I^\alpha t^k \\ &= i \sum_{k=0}^{i} (-1)^{i-k} \frac{(i + k - 1)!\, 2^{2k} k!}{(i - k)!(2k)!\, \Gamma(k + \alpha + 1)} t^{k+\alpha}, \quad i = 0, 1, \ldots, n. \end{aligned} \tag{6.39}$$

Now, approximate $t^{k+\alpha}$ by $n + 1$ terms of shifted Chebyshev series, we have

$$t^{k+\alpha} = \sum_{j=0}^{n} c_{kj}\, T_j^*(t), \tag{6.40}$$

where c_{kj} is given from Eq. (6.35) with $y(t) = t^{k+\alpha}$, that is

$$c_{kj} = \begin{cases} \frac{1}{\sqrt{\pi}} \frac{\Gamma(k+\alpha+\frac{1}{2})}{\Gamma(k+\alpha+1)}, & j = 0, \\ \frac{j}{\sqrt{\pi}} \sum_{r=0}^{j} (-1)^{j-r} \frac{(j+r-1)! 2^{2r+1} \Gamma(k+r+\alpha+\frac{1}{2})}{(j-r)!(2r)!\Gamma(k+\alpha+r+1)}, & j = 1, \dots, n. \end{cases} \tag{6.41}$$

In virtue of Eqs. (6.39) and (6.40), we get

$$\begin{aligned} I^{\alpha} T_i^*(t) &= i \sum_{k=0}^{i} (-1)^{i-k} \frac{(i+k-1)! 2^{2k}}{(i-k)!(2k)! p(k+\alpha+1)} \Big(\sum_{j=0}^{n} c_{kj} T_j^*(t) \Big) \\ &= \sum_{j=0}^{n} \sum_{k=0}^{i} \epsilon_{ijk} \, T_j^*(x) = \sum_{j=0}^{n} \Omega_{\alpha}(i,j) T_j^*(t), \end{aligned} \tag{6.42}$$

where $\Omega_{\alpha}(i,j) = \sum_{k=0}^{i} \epsilon_{ijk}$, and

$$\epsilon_{ijk} = \begin{cases} \dfrac{i(-1)^{i-k}(i+k-1)! 2^{2k} k! \Gamma(k+\alpha+\frac{1}{2})}{(i-k)!(2k)!\sqrt{\pi}(\Gamma(k+\alpha+1))^2}, & j = 0 \\ \dfrac{(-1)^{i-k} ij(i+k-1)! 2^{2k+1} k!}{(i-k)!(2k)!\Gamma(k+\alpha+1)\sqrt{\pi}} \\ \times \displaystyle\sum_{r=0}^{j} \dfrac{(-1)^{i-r}(i+r-1)! 2^{2r} \Gamma(k+r+\alpha+\frac{1}{2})}{(j-r)!(2r)!\Gamma(k+r+\alpha+1)}, & j = 1, \dots, n. \end{cases}$$

Accordingly, Eq. (6.42) can be written in a vector form as follows:

$$I^{\alpha} T_i^*(t) \simeq [\Omega_{\alpha}(i,0) \quad \Omega_{\alpha}(i,1) \quad \dots \quad \Omega_{\alpha}(i,n)] \, \Phi(t), \qquad i = 0, 1, \dots, n. \tag{6.43}$$

Equation (6.43) leads to the desired result and the operational matrix of fractional integration of order α is as

$$P^{\alpha} = \begin{bmatrix} \Omega_{\alpha}(0,0) & \Omega_{\alpha}(0,1) & \dots & \Omega_{\alpha}(0,n) \\ \Omega_{\alpha}(1,0) & \Omega_{\alpha}(1,1) & \dots & \Omega_{\alpha}(1,n) \\ \vdots & \vdots & \dots & \vdots \\ \Omega_{\alpha}(i,0) & \Omega_{\alpha}(i,1) & \dots & \Omega_{\alpha}(i,n) \\ \vdots & \vdots & \dots & \vdots \\ \Omega_{\alpha}(n,0) & \Omega_{\alpha}(n,1) & \dots & \Omega_{\alpha}(n,n) \end{bmatrix}. \tag{6.44}$$

6.2.2 Applications

Example 6.3 *We consider the following initial value problem [10]*

$$D^{\frac{3}{2}} y(t) + 3y(t) = 3t^3 + \frac{8}{\Gamma(0.5)} t^{1.5}, \quad t \in [0,1], \tag{6.45}$$

$$y(0) = y'(0) = 0, \tag{6.46}$$

whose exact solution is given by $y(t) = t^3$.

Here, we apply the technique described in Section 6.1.2 based on the shifted Chebyshev polynomials with $n = 3$. By applying the Riemann-Liouville integral of order $\frac{3}{2}$ on Eq. (6.45), we may approximate solution and the right-hand side in the forms

$$y(t) = \sum_{i=0}^{3} c_i T_i^*(t) = C^T \Phi(t), \qquad g(t) = \sum_{i=0}^{3} g_i T_i^*(t) = G^T \Phi(t), \tag{6.47}$$

According to Eq. (6.44), the operational matrix of fractional integration of order $\frac{3}{2}$ is as following

$$P^{\frac{3}{2}} = \frac{4}{\pi^{\frac{3}{2}}} \begin{bmatrix} \frac{4}{9} & \frac{8}{15} & \frac{8}{105} & \frac{-8}{945} \\ \frac{-4}{25} & \frac{-8}{63} & \frac{8}{135} & \frac{8}{385} \\ \frac{-20}{147} & \frac{-152}{675} & \frac{-248}{3465} & \frac{136}{4095} \\ \frac{188}{2025} & \frac{136}{1617} & \frac{-56}{975} & \frac{-664}{17325} \end{bmatrix},$$

and $G = [g_0, g_1, g_2, g_3]^T$. Therefore using the residual $R_n(t)$ and the typical tau method we obtain

$$\frac{96}{105\pi^{\frac{3}{2}}} c_0 + \frac{96}{135\pi^{\frac{3}{2}}} c_1 + \frac{2976}{3465\pi^{\frac{3}{2}}} c_2 - \frac{672}{975\pi^{\frac{3}{2}}} c_3 + c_2 - g_2 = 0, \tag{6.48}$$

$$-\frac{96}{945\pi^{\frac{3}{2}}} c_0 + \frac{96}{385\pi^{\frac{3}{2}}} c_1 + \frac{1632}{4095\pi^{\frac{3}{2}}} c_2 - \frac{7968}{17325\pi^{\frac{3}{2}}} c_3 + c_3 - g_3 = 0. \tag{6.49}$$

Using the initial conditions (6.46), we have

$$C^T \Phi(0) = c_0 - c_1 + c_2 - c_3 = 0, \tag{6.50}$$

$$C^T D^{(1)} \Phi(0) = 2c_1 - 8c_2 + 18c_3 = 0. \tag{6.51}$$

Finally by solving Eqs. (6.48)–(6.51), we get

$$c_0 = \frac{5}{16}, \quad c_1 = \frac{15}{32}, \quad c_2 = \frac{3}{16}, \quad c_3 = \frac{1}{32}. \tag{6.52}$$

Thus we can write

$$y(t) = \sum_{i=0}^{3} c_i T_i^*(t) = t^3. \tag{6.53}$$

which is the exact solution.

Example 6.4 *Consider the following initial value problem [10]*

$$D^2 y(t) - 2Dy(t) + D^{\frac{1}{2}} y(t) + y(t) = t^3 - 6t^2 + 6t + \frac{16}{5\sqrt{\pi}} t^{2.5}, \quad t \in [0, 1],$$
$$y(0) = y'(0) = 0, \tag{6.54}$$

whose exact solution is given by $y(t) = t^3$.

Now, we can apply the technique described in Example 6.3 with $n = 3$, and the four unknown coefficients will be in the form

$$c_0 = \frac{5}{16}, \quad c_1 = \frac{15}{32}, \quad c_2 = \frac{3}{16}, \quad c_3 = \frac{1}{16}.$$

Thus we can write, $y(t) = \sum_{i=0}^{3} c_i T_i^*(t) = t^3$, which is the exact solution.

6.3 BERNSTEIN POLYNOMIALS

The Bernstein polynomials (Bps) of degree n on $[0, 1]$ are defined as follows [47,119]:

$$B_{i,n}(t) = \binom{n}{i} t^i (1-t)^{n-i}, \qquad 0 \leq i \leq n.$$

Bernstein polynomials defined above form a complete basis over the interval $[0, 1]$. There are $n + 1$ nth-degree polynomials. For convenience, we set $B_{i,n}(x) = 0$, if $i < 0$ or $i > n$. A recursive definition can also be used to generate the Bernstein polynomials over $[0, 1]$ so that the ith nth-degree Bernstein polynomials can be written

$$B_{i,n}(t) = -xB_{i,n-1}(t) + tB_{i-1,n-1}(t).$$

It can be easily shown that each of the Bernstein polynomials is positive and also the sum of all the Bernstein polynomials is unity for all real $x \in [0, 1]$, i.e., $\sum_{i=0}^{n} B_{i,n}(t) = 1$. It is easy to show that any given polynomial of degree n can be expanded in terms of linear combination of the basis functions.

By using the binomial expansion of $(1-t)^{n-i}$, one can show that

$$\begin{aligned} B_{i,n}(t) &= \binom{n}{i} t^i (1-t)^{n-i} = \binom{n}{i} t^i \left(\sum_{k=0}^{n-i} (-1)^k \binom{n-i}{k} t^k \right) \\ &= \sum_{k=0}^{n-i} (-1)^k \binom{n}{i} \binom{n-i}{k} t^{i+k}, \qquad i = 0, 1, \ldots, n. \end{aligned} \tag{6.55}$$

A function $y(t)$, square integrable in $[0, 1]$, can be expressed in terms of Bernstein polynomials as

$$y(t) = \sum_{i=0}^{\infty} c_i B_{i,n}(t),$$

In practice, only the first $(n + 1)$-terms of Bernstein polynomials are considered. Therefore, one has

$$y_n(t) \simeq \sum_{i=0}^{n} c_i B_{i,n}(t) = C^T \Phi(t), \tag{6.56}$$

where the vectors C and $\Phi(t)$ are given by

$$C = [c_0, c_1, \ldots, c_n]^T, \qquad \Phi(t) = [B_{0,n}(t), B_{1,n}(t), \ldots, B_{n,n}(t)]^T, \tag{6.57}$$

and C is given by

$$C^T = \left(\int_0^1 y(t)\Phi(t)dt \right) Q^{-1}, \tag{6.58}$$

where Q is an $(n+1) \times (n+1)$ matrix and is defined as

$$Q = \langle \Phi, \Phi \rangle = \int_0^1 \Phi(t)\Phi(t)^T dt.$$

Note: The vector $\Phi(t)$ defined in Eq. (6.57) can be written as [105]

$$\Phi(t) = AT_n(t), \tag{6.59}$$

where $T_n(t) = [1, t, \ldots, t^n]^T$ and $A = (\lambda_{i,j})_{i,j=0}^n$ is given in the following form:

$$\lambda_{i,j} = \begin{cases} (-1)^{j-i}\binom{n}{i}\binom{n-i}{j-i}, & i \leq j, \\ 0, & i > j, \end{cases} \qquad i, j = 0, 1, \ldots, m.$$

6.3.1 Operational Matrix of Fractional Integration

In this subsection, we want to investigate an operational matrix of fractional integration for Bps. Therefore, by fractional integration of the vector $\phi(t)$ as below, we get [65, 69, 102]:

$$I^\alpha \Phi(t) \simeq P^\alpha \Phi(t), \tag{6.60}$$

where P^α is the $(n+1) \times (n+1)$ Riemann-Liouville fractional operational matrix of integration for BPs. Instead of using $\Phi(t)$ we can substitute $AT_n(t)$, in consequence we get:

$$\begin{aligned} I^\alpha \Phi(t) &= I^\alpha AT_n(t) = AI^\alpha T_n(t) = A\left[I^\alpha 1, I^\alpha t, \ldots, I^\alpha t^n\right]^T \\ &= A\left[\frac{0!}{\Gamma(\alpha+1)}t^\alpha, \frac{1!}{\Gamma(\alpha+2)}t^{\alpha+1}, \ldots, \frac{n!}{\Gamma(\alpha+n+1)}t^{\alpha+n}\right]^T = A\Theta\overline{T}_n(t), \end{aligned}$$

where Θ, being an $(n+1) \times (n+1)$ matrix, and $\overline{T}_n(t)$ are given by

$$\Theta_{i,j} = \begin{cases} \frac{i!}{\Gamma(\alpha+i+1)}, & i = j, \\ \\ 0, & i \neq j. \end{cases} \qquad i, j = 0 \ldots, n,$$

$$\overline{T}_n = \left[t^\alpha, t^{\alpha+1}, \ldots, t^{\alpha+n}\right]^T.$$

Now, approximate $t^{l+\alpha}$ by $n+1$ terms of Bernstein polynomials, we have

$$t^{\alpha+l} = w_l^T \Phi(t), \qquad l = 0, \ldots, n.$$

Therefore we have

$$\begin{aligned} w_l &= Q^{-1}\left(\int_0^1 t^{\alpha+l}\Phi(t)dt\right) \\ &= Q^{-1}\left[\int_0^1 t^{\alpha+l}B_{0,n}(t)dt, \int_0^1 t^{\alpha+l}B_{1,n}(t)dt, \ldots, \int_0^1 t^{\alpha+l}B_{n,n}(t)dt\right]^T = Q^{-1}\overline{w}_l, \end{aligned}$$

where $\overline{w}_l = [\overline{w}_{l,0}, \overline{w}_{l,1}, \ldots, \overline{w}_{l,n}]^T$ and

$$\overline{w}_{l,k} = \int_0^1 t^{\alpha+l}B_{k,n}(t)dt = \frac{n!\Gamma(l+k+\alpha+1)}{k!\Gamma(l+n+\alpha+2)}, \qquad l, k = 0, 1, \ldots, n,$$

where $w = [w_0, w_1, \ldots, w_n]^T$ is an $(n+1) \times (n+1)$ matrix that has vector $Q^{-1}\overline{w}_l$ for the i-th columns. Therefore, we can write

$$I^\alpha \Phi(t) \simeq P^\alpha \Phi(t) = A\Theta w^T \Phi(t), \tag{6.61}$$

where $P^\alpha = A\Theta w^T$ is called the operational matrix of fractional integral.

6.3.2 Operational Matrix of Ordinary Derivative

The first-order derivative of the vector $\Phi(t)$ can be expressed by the Bps as follows [105]:

$$\frac{d}{dt}\Phi(t) \simeq \mathbf{D}^1\, \Phi(t), \tag{6.62}$$

where $\mathbf{D}^1_{(n+1)\times(n+1)}$ is the derivative operational matrix (OM) of integer order. Since $\Phi(t) = AT_n(t)$, we can write

$$\begin{aligned} \frac{d}{dt}\phi(t) &= A\frac{d}{dt}T_n(t)A\begin{bmatrix} 0 \\ 1 \\ 2t \\ \vdots \\ nt^{n-1} \end{bmatrix} \\ &= A\begin{bmatrix} 0 & 0 & \ldots & 0 \\ 1 & 0 & \ldots & 0 \\ 0 & 2 & \ldots & 0 \\ \vdots & \vdots & \ddots & \vdots \\ 0 & 0 & \ldots & m \end{bmatrix}\begin{bmatrix} 1 \\ t \\ t^2 \\ \vdots \\ t^{n-1} \end{bmatrix} = A\Sigma T_n^*(t), \end{aligned}$$

where

$$\Sigma = \begin{bmatrix} 0 & 0 & \dots & 0 \\ 1 & 0 & \dots & 0 \\ 0 & 2 & \dots & 0 \\ \vdots & \vdots & \ddots & \vdots \\ 0 & 0 & \dots & n \end{bmatrix}, \quad \text{and} \quad T_n^*(t) = \begin{bmatrix} 1 \\ t \\ t^2 \\ \vdots \\ t^{n-1} \end{bmatrix}.$$

Now, we can compute the vector $T_n^*(t)$ in terms of Bps as

$$T_n^*(t) = B^*\Phi(t),$$

so that,

$$B^* = \begin{bmatrix} A_1^{-1} \\ A_2^{-1} \\ \vdots \\ A_n^{-1} \end{bmatrix},$$

in which A_i^{-1} is the ith row of A^{-1}, $i = 1, 2, \dots, n$. Thus $\frac{d}{dt}\Phi(t) = A\Sigma B^*\Phi(t)$. Hence the OM of derivative is

$$\mathbf{D}^1 = A\Sigma B^*. \tag{6.63}$$

By using Eq. (6.62), it is clear that $\frac{d^n\Phi(t)}{dt^n} = \mathbf{D}^n\Phi(t)$, where $\mathbf{D}^n$ is nth power of $\mathbf{D}^1$.

6.3.3 Operational Matrix of Fractional Derivative

The Caputo fractional derivative of the vector $\psi(t)$ can be expressed by Ref. [105]

$$D^\alpha\Phi(t) \simeq \mathbf{D}^\alpha\Phi(t), \tag{6.64}$$

where D^α represents the $(n+1) \times (n+1)$ operational matrix of the fractional derivative. By using Eq. (6.59), we achieve

$$D^\alpha\Phi(t) = AD^\alpha T_n(t) = \bar{\Upsilon}\bar{T},$$

where $\overline{\Upsilon} = (\overline{\Upsilon_{k,\iota}})_{k,\iota=1}^{n+1}$ is an $(n+1)$ order diagonal matrix as

$$\overline{\Upsilon}_{k+1,\iota+1} = \begin{cases} \dfrac{\Gamma(\iota+1)}{\Gamma(\iota+1-\alpha)}, & k, \iota = \lceil\alpha\rceil, \dots, n, \quad k = \iota, \\ 0, & \text{otherwise}, \end{cases}$$

and $\overline{T}_{(n+1)\times 1}$ is defined by

$$\overline{T}_{k+1} = \begin{cases} 0, & k = 0, \dots, \lceil\alpha\rceil - 1, \\ t^{k-\alpha}, & k = \lceil\alpha\rceil, \dots, n. \end{cases}$$

Now we expand $\overline{T}$ by terms of the Bernstein basis as $\overline{T} \simeq \Pi^T \Phi(t)$, where

$$\begin{aligned}\Pi &= [\Pi_0, \Pi_1, \ldots, \Pi_n],\\ \Pi_k &= \Psi^{-1}\hat{\Pi}_k,\\ \hat{\Pi}_k &= [\hat{\Pi}_{k,0}, \hat{\Pi}_{k,2}, \ldots, \hat{\Pi}_{k,n}]^T,\end{aligned}$$

and

$$\hat{\Pi}_{k,\iota} = \int_0^1 t^{k-\theta}\mathcal{B}_{k,\iota}(t)dt = \frac{n!\Gamma(k+\iota-\alpha+1)}{\iota!\Gamma(k+m-\alpha+2)}, \qquad k,\iota = 0,1,\ldots,n.$$

Finally, we obtain $D^\alpha \Phi(t) \simeq \mathbf{D}^\alpha \Phi(t)$, where

$$\mathbf{D}^\alpha = A\Upsilon\Pi^T. \tag{6.65}$$

is called the operational matrix of the fractional derivative.

6.3.4 Operational Matrix of Product

It is always necessary to assess the product of $\phi(t)$ and $\phi(t)^T$, which is called the product matrix for the Bps basis. The operational matrices for the product $\hat{C}$ are given by Refs. [65, 66, 69, 102]:

$$C^T\Phi(t)\Phi(t)^T \simeq \Phi(t)^T\hat{C}, \tag{6.66}$$

where $\hat{C}$ is an $(n+1)\times(n+1)$ matrix. So we have

$$\begin{aligned}C^T\Phi(t)\Phi(t)^T &= C^T\Phi(t)\left(T_n(t)^T A^T\right) = [C^T\phi(t), t(C^T\Phi(t)), \ldots, t^n(C^T\Phi(t))]A^T\\ &= \left[\sum_{i=0}^n c_iB_{i,n}(t), \sum_{i=0}^n c_itB_{i,n}(t), \ldots, \sum_{i=0}^n c_it^nB_{i,n}(t)\right].\end{aligned} \tag{6.67}$$

Now, we approximate all functions $t^kB_{i,n}(t)$ in terms of $\{B_{i,n}\}_{i=0}^n$ for $i,k = 0,1,\ldots,n$, as following

$$t^kB_{i,m}(t) = e_{k,i}^T\phi_m(t).$$

that $e_{k,i} = \left[e_{k,i}^0, e_{k,i}^1, \ldots, e_{k,i}^n\right]^T$, then we obtain the components of the vector of $e_{k,i}$

$$\begin{aligned}e_{k,i} &= Q^{-1}\left(\int_0^1 t^kB_{i,n}(t)\phi(t)dt\right)\\ &= Q^{-1}\left[\int_0^1 t^kB_{i,n}(t)B_{0,n}(t)dt, \int_0^1 t^kB_{i,n}(t)B_{1,n}(t)dt, \ldots, \int_0^1 t^kB_{i,n}(t)B_{n,n}(t)dt\right]^T\\ &= \frac{Q^{-1}}{2n+k+1}\left[\frac{\binom{n}{0}}{\binom{2n+k}{i+k}}, \frac{\binom{n}{1}}{\binom{2n+k}{i+k+1}}, \ldots, \frac{\binom{n}{n}}{\binom{2n+k}{i+k+n}}\right]^T, \qquad i,k = 0,1,\ldots,n.\end{aligned}$$

Thus, we obtain finally

$$\begin{aligned}\sum_{i=0}^{n} c_i t^k B_{i,n}(t) &= \sum_{i=0}^{n} c_i \left(\sum_{j=0}^{n} e_{k,i}^{j} B_{j,n}(t)\right) = \sum_{j=0}^{n} B_{j,n}(t)\left(\sum_{i=0}^{n} c_i e_{k,i}^{j}\right) \\ &= \Phi_n(t)^T \left[\sum_{i=0}^{n} c_i e_{k,i}^{0}, \sum_{i=0}^{n} c_i e_{k,i}^{1}, \ldots, \sum_{i=0}^{n} c_i e_{k,i}^{n}\right]^T \\ &= \Phi(t)^T \left[e_{k,0}, e_{k,1}, \ldots, e_{k,n}\right] C = \Phi(t)^T V_{k+1} C, \end{aligned} \quad (6.68)$$

where $V_{k+1}(k = 0, 1 \ldots, n)$ is an $(n+1) \times (n+1)$ matrix that has vectors $e_{k,i}(i = 0, 1, \ldots, n)$ given, for each columns. If we choose an $(n+1) \times (n+1)$ matrix $\bar{C} = [V_1 c, V_1 c, \ldots, V_{n+1} c]$, from Eqs. (6.67) and (6.68) we can write

$$C^T \Phi(t) \Phi(t)^T \simeq \Phi(t)^T \bar{C} A^T, \quad (6.69)$$

and therefore we obtain the operational matrix of product, $\hat{C} = \bar{C} A^T$.

Corollary 6.1 *If $y(t) = C^T \phi(t)$, consequently we can get the approximate function for $y^k(t)$, using BPs as follows:*

$$y^k(t) \simeq \phi(t)^T \tilde{C}_k, \quad (6.70)$$

where $\tilde{C}_k = \hat{C}^{k-1} C$ and $\hat{C}$ is $(n+1) \times (n+1)$ operational matrix of product using BPs.

6.3.5 Applications

In this section, in order to show the high importance of Bernstein operational matrices, we apply those to solve multi-order fractional differential equation and nonlinear fractional differential equation.

6.3.5.1 *Method 1*

Here, a numerical method based on the operational martrix of fractional derivative is applied to solving the multi-term FDE [98]

$$F\left(t, y(t), D^{\beta_1} y(t), \ldots, D^{\beta_k} y(t)\right) = 0, \quad (6.71)$$

with initial conditions

$$y^{(i)}(0) = d_i, \quad i = 0, 1, \ldots, p, \quad (6.72)$$

where $0 \leq p < \max\{\beta_i,\ i = 1, \ldots, k\} \leq p + 1$. It should be noted that F can be nonlinear in general. In order to use Bernstein polynomials for this problem, we approximate $y(t)$ by Bernstein polynomials as

$$y(t) \simeq \sum_{i=0}^{n} c_i B_{i,n}(t) = C^T \Phi(t), \quad (6.73)$$

where vector $= [c_0, \ldots, c_n]^T$ is unknown vector. By using the operational matrix of fractional derivative (6.64) and (6.73), we have

$$D^{\beta_j} y(t) \simeq C^T D^{\beta_j} \Phi(t) = C^T \mathbf{D}^{\beta_j} \Phi(t), \quad j = 1, \ldots, k. \tag{6.74}$$

By substituting Eqs. (6.73) and (6.74) in Eq. (6.71), we have

$$F\left(t, C^T\Phi(t), C^T\mathbf{D}^{\beta_j}\Phi(t), \ldots, C^T\mathbf{D}^{\beta_k}\Phi(t)\right) = 0, \tag{6.75}$$

Similarly, substituting Eq. (6.73) in (6.72) yields

$$C^T\mathbf{D}^{\beta_j}\Phi(0) = d_i, \quad i = 0, 1, \ldots, p, \tag{6.76}$$

To find the solution $y(t)$, we first collocate Eq. (6.75) at $(n-p)$ points. For suitable collocation points, we use

$$t_i = \frac{1}{2}\left(\cos\left(\frac{i\pi}{n}\right) + 1\right), \quad i = 1, \ldots, n-p.$$

These equations together with Eq. (6.76) generate $n+1$ algebraic equations which can be solved to find c_i, $i = 0, \ldots, n$. Consequently the unknown function $y(t)$ given in Eq. (6.73) can be calculated.

6.3.5.2 Method 2

Here, a numerical method based on operational matrix of fractional integral and operational matrix of product is applied to solving the nonlinear FDE as:

$$D^\alpha y(t) = \sum_{i=1}^{k} \mu_i y^i(t) + \mu_{k+1} y(t) + f(t), \tag{6.77}$$

with initial conditions

$$y^{(i)}(0) = d_i, \quad i = 0, \ldots, m-1, \tag{6.78}$$

where $\mu_i (i = 1, 2, \ldots, k)$ are real constant coefficients and also $m-1 < \alpha < m$. The values of $d_i (i = 0, \ldots, m-1)$ describe the initial state of $y(t)$ and $f(t)$ is a given source function.

Now, we approximate $D^\alpha y(t)$ and $f(t)$ by means of the Bernstein polynomials:

$$D^\alpha y(t) \simeq \sum_{i=0}^{n} c_i B_{i,n}(t) = C^T\Phi(t) \tag{6.79}$$

$$f(t) \simeq \sum_{i=0}^{n} f_i B_{i,n}(t) = F^T\Phi(t), \tag{6.80}$$

where

$$C = [c_0, c_1, \ldots, c_n]^T$$
$$F = [f_0, f_1, \ldots, f_n]^T$$

Applying the Riemann-Liouville fractional integration on both sides of Eq. (6.79), we get:

$$\begin{aligned} y(t) &= C^T I^\alpha \Phi(t) + \sum_{i=0}^{n-1} \frac{y^{(i)}(0)}{i!} t^i = C^T P^\alpha \Phi(t) + Y^T \Phi(t) \\ &= (C^T P^\alpha + Y^T)\Phi(t) = M_\alpha \Phi(t), \end{aligned} \tag{6.81}$$

where $M_\alpha = (C^T P^\alpha + Y^T)$ and P^α is the $(n+1) \times (n+1)$ operational matrix of fractional integration of order α based on Bernstein polynomials. Also

$$\sum_{i=0}^{n-1} \frac{y^{(i)}(0)}{i!} t^i = Y^T \Phi(t),$$

where $Y^T = [y_0, \ldots, y_n]$ is known coefficient. For nonlinear term, based on operational matrix of product (6.66), we get

$$y^2(t) \simeq M_\alpha^T \Phi(t) \Phi^T(t) M_\alpha \simeq \Phi(t)^T \hat{M}_\alpha M_\alpha \simeq M_\alpha^T \hat{M_\alpha}^T \Phi(t), \tag{6.82}$$

$$y^k(t) \simeq \Phi(t)^T \tilde{M}_k, \tag{6.83}$$

where $\tilde{M}_k = \hat{M_\alpha}^{k-1} M_\alpha$ and $\hat{M}_\alpha$ is $(n+1) \times (n+1)$ operational matrix of product using BPs. By substituting Eqs. (6.79)–(6.83) in Eq. (6.77), we obtain

$$C^T \Phi(t) = \sum_{i=1}^{k} \mu_i \tilde{M}_{i-1}^T \Phi(t) + \mu_{k+1} M_\alpha \Phi(t) + F^T \Phi(t),$$

or

$$C^T - \sum_{i=1}^{k} \mu_i \tilde{M}_{i-1}^T - \mu_{k+1} M_\alpha - F^T = 0. \tag{6.84}$$

Finally, the vector C can be obtained by solving the system (6.84). Consequently determine the approximate value of $y(t)$ can be calculated from Eq. (6.79).

6.3.5.3 *Method 3*

In this section, we provide an optimization method to solving the multi-term FDE [105]

$$D^\alpha y(t) = F\left(t, y(t), D^{\beta_1} y(t), \ldots, D^{\beta_k} y(t)\right), \quad 0 \le t \le 1, \tag{6.85}$$

with initial conditions

$$y^{(i)}(0) = d_i, \quad i = 0, 1, \ldots, m-1, \tag{6.86}$$

where $m - 1 < \alpha < m$, $0 < \beta_1 < \beta_2 < \cdots < \beta_k < \alpha$. It should be noted that F can be nonlinear in general. To this end, we approximate $y(t)$ in terms of Bps as:

$$y(t) \simeq C^T \Phi(t), \tag{6.87}$$

where $C^T = [c_0, c_1, \dots, c_n]$ is an unknown $1 \times (n+1)$ vector and $\Phi(t)$ is defined in Eq. (6.57). By employing the fractional derivative operator of order α and $\beta_j, (j = 1, \dots, k)$ on both sides of Eq. (6.87) and using the operational matrix of fractional derivative, we get

$$D^\alpha y(t) \simeq D^\alpha C^T \Phi(t) \simeq C^T D^\alpha \Phi(t) \simeq C^T \mathbf{D}^\alpha \Phi(t), \tag{6.88}$$
$$D^{\beta_j} y(t) \simeq C^T D^{\beta_j} \Phi(t) \simeq C^T \mathbf{D}^{\beta_j} \Phi(t), \quad j = 1, \dots, k. \tag{6.89}$$

Substituting Eqs. (6.87)–(6.89) in Eq. (6.85), we reach

$$C^T \mathbf{D}^\alpha \Phi(t) - F\left(t, C^T \Phi(t), C^T \mathbf{D}^{\beta_j} \Phi(t), \dots, C^T \mathbf{D}^{\beta_k} \Phi(t)\right) \triangleq R(t). \tag{6.90}$$

To reach suitable values for the unknown vector V and the approximate solution for the given problem, the 2-norm of Eq. (6.90) is defined as

$$M(V) = \int_0^1 R^2(t)dt.$$

To achieve the optimal values for the unknown vector V, we introduce

$$\min \quad J[V] = M(V), \tag{6.91}$$

subject to

$$\begin{aligned}
\Delta_0 &\triangleq C^T \Phi(0) - d_0 = 0, \\
\Delta_1 &\triangleq C^T \mathbf{D}^1 \Phi(0) - d_1 = 0, \\
&\vdots \\
\Delta_{n-1} &\triangleq C^T \mathbf{D}^{n-1} \Phi(0) - d_{n-1} = 0.
\end{aligned}$$

Using the method of Lagrange multiplier, we have

$$J^*[C, \lambda] = J[C] + \lambda \Delta, \tag{6.92}$$

where $\lambda = [\lambda_0, \lambda_1, \dots, \lambda_{n-1}]$ and Δ is the known column vector

$$\Delta = \begin{bmatrix} \Delta_0 \\ \Delta_1 \\ \vdots \\ \Delta_{n-1} \end{bmatrix}.$$

According to differential calculus, the fundamental conditions for the optimality are

$$\frac{\partial J^*}{\partial C} = 0, \qquad \frac{\partial J^*}{\partial \lambda} = 0. \tag{6.93}$$

The values of C and λ can be gained by solving Eq. (6.93). Then, by inserting in Eq. (6.87), the approximate solution of the consideration model is achieved.

Example 6.5 *Consider the inhomogeneous Bagley-Torvik equation [98]*

$$D^2y(t) + D^{\frac{3}{2}}y(t) + y(t) = 1 + t, \tag{6.94}$$

with initial condition

$$y(0) = 0, \quad y'(0) = 1. \tag{6.95}$$

that the exact solution is $y(t) = 1 + t$.

By applying the method 1 with $n = 2$, we approximate solution as

$$y(t) \simeq \sum_{i=0}^{2} c_i B_{i,2}(t) = C^T \Phi(t). \tag{6.96}$$

Here, we have

$$\mathbf{D}^1 = \begin{bmatrix} -2 & -1 & 0 \\ 2 & 0 & -2 \\ 0 & 1 & 2 \end{bmatrix}, \quad \mathbf{D}^2 = \begin{bmatrix} 2 & 2 & 2 \\ -4 & -4 & -4 \\ 2 & 2 & 2 \end{bmatrix},$$

$$\mathbf{D}^{\frac{3}{2}} = \left(\frac{8}{35\sqrt{\pi}}\right) \begin{bmatrix} 3 & 15 & 17 \\ -6 & -30 & -34 \\ 3 & 15 & 17 \end{bmatrix},$$

By substituting these equations in Eq. (6.94), we get

$$C^T \mathbf{D}^2 \Phi(t) + C^T \mathbf{D}^{\frac{3}{2}} \Phi(t) + C^T \Phi(t) = 1 + t, \tag{6.97}$$

By collocating Eq. (6.97) at $\frac{1}{2}$, we obtain

$$\frac{9}{4}(c_0 + c_2) - \frac{7}{2}c_1 + \frac{20}{7\sqrt{\pi}}(c_0 + 2c_1 + c_2) - \frac{3}{2} = 0. \tag{6.98}$$

Now, by using the initial conditions, we have

$$c_0 - 1 = 0, \tag{6.99}$$

$$-2c_0 + 2c_1 - 1 = 0. \tag{6.100}$$

Finally by solving Eqs. (6.98)–(6.100), we get

$$c_0 = 1, \quad c_1 = \frac{3}{2}, \quad c_2 = 2. \tag{6.101}$$

Thus

$$y(t) = \begin{bmatrix} 1 & \frac{3}{2} & 2 \end{bmatrix} \begin{bmatrix} (1-t)^2 \\ 2t(1-t) \\ t^2 \end{bmatrix} = 1 + t, \tag{6.102}$$

which is the exact solution.

Example 6.6 *We solve below the following fractional Riccati differential equation*

$$D^{\alpha}y(t) = -y^2(t) + 1, \qquad 0 < \alpha \leq 1, \tag{6.103}$$

with initial condition

$$y(0) = 0. \tag{6.104}$$

The exact solution, when $\alpha = 1$, is

$$y(t) = \frac{e^{2t} - 1}{e^{2t} + 1}. \tag{6.105}$$

We applied method 2 to solve Eq. (6.103). Table 6.1 shows the absolute errors of $y(t)$ for $n = 6$ and different values of $\alpha = 0.96, 0.98, 1$. Definitely, as α approaches 1, the approximate values of $y(t)$ will converge to the exact solutions. In Table 6.2, the absolute errors of $y(t)$ shows for $\alpha = 1$ and different values of n. The approximate

TABLE 6.1 Absolute Errors of $y(t)$ for $n = 6$ and Different Values of α for Example 6.6

t	$\alpha = 1$	$\alpha = 0.98$	$\alpha = 0.96$
0	6.32669×10^{-6}	4.62577×10^{-4}	1.00175×10^{-3}
0.1	1.874×10^{-6}	5.51534×10^{-3}	1.13019×10^{-2}
0.2	2.05157×10^{-6}	7.8834×10^{-3}	1.59905×10^{-2}
0.3	3.40078×10^{-7}	8.72049×10^{-3}	1.75444×10^{-2}
0.4	1.9600×10^{-6}	8.58365×10^{-3}	1.71425×10^{-2}
0.5	3.85062×10^{-7}	7.72916×10^{-3}	1.53192×10^{-2}
0.6	1.91429×10^{-6}	6.32137×10^{-3}	1.24157×10^{-2}
0.7	4.06158×10^{-7}	4.5515×10^{-3}	8.82564×10^{-3}
0.8	2.12521×10^{-6}	2.65545×10^{-3}	5.02479×10^{-3}
0.9	1.36592×10^{-6}	8.16445×10^{-4}	1.37036×10^{-3}
1	6.32669×10^{-6}	1.06106×10^{-3}	2.34375×10^{-3}

TABLE 6.2 Absolute Errors of $y(t)$ for $\alpha = 1$ and Different Values of n for Example 6.6

t	$n = 2$	$n = 5$	$n = 10$
0	5.41333×10^{-3}	5.06438×10^{-5}	5.1331×10^{-9}
0.1	8.31268×10^{-4}	1.90542×10^{-5}	1.43088×10^{-9}
0.2	2.66311×10^{-3}	1.11372×10^{-5}	1.13828×10^{-9}
0.3	1.89155×10^{-3}	1.30721×10^{-5}	1.27298×10^{-9}
0.4	4.75147×10^{-5}	7.61542×10^{-6}	1.0664×10^{-9}
0.5	1.70179×10^{-3}	1.54903×10^{-5}	2.61128×10^{-10}
0.6	2.58874×10^{-3}	6.02423×10^{-8}	7.13941×10^{-10}
0.7	2.23492×10^{-3}	1.64551×10^{-5}	1.25679×10^{-9}
0.8	6.053×10^{-4}	5.00047×10^{-6}	1.33421×10^{-9}
0.9	2.05878×10^{-3}	2.07919×10^{-5}	1.56084×10^{-9}
1	5.31426×10^{-3}	5.06432×10^{-5}	5.1331×10^{-9}

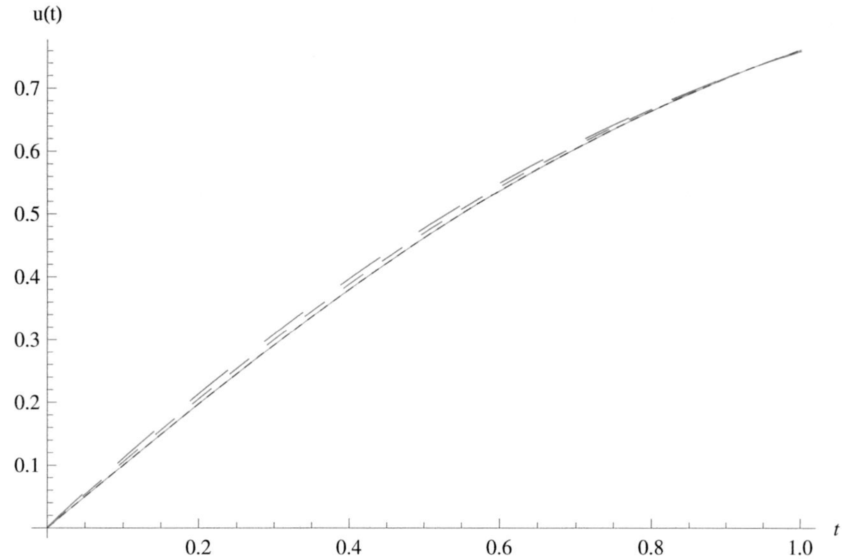

Figure 6.1 Comparison of the exact solution (blue line) with the approximate solution $y(t)$ for $n = 6$ and $\alpha = 1$ (dotted), $\alpha = 0.98$ (dashed line), $\alpha = 0.96$ (long dashed line) for Example 6.6.

values of $y(t)$ converge to the exact solutions for a sufficient large values of n. Figure 6.1 shows the comparison between the exact solution together with the approximate solution $y(t)$ for $n = 6$ and different values of α.

Example 6.7 *We consider the following fractional Duffing equation [105]*

$$D^{\alpha}y(t) + 5y^{''}(t) + 4y(t) - \frac{1}{6}y^{3}(t) = 0, \qquad 3 < \alpha \leq 4. \tag{6.106}$$

with the conditions

$$y(0) = 0, \qquad y^{'}(0) = 1.91103, \qquad y''(0) = 0, \qquad y'''(0) = -1.15874, \tag{6.107}$$

when $\alpha = 4$, *the exact solution is the following form*

$$y(t) = 2.1906\sin(0.9t) - 0.02247\sin(2.7t) + 0.000045\sin(4.5t).$$

We apply method 3 described in this section. Table 6.3 presents the absolute errors for $\alpha = 4$ and $n = 5, 8$. Also the absolute errors for $n = 6$ and the different values of $\alpha = 3.97, 3.99, 4$ are shown in Table 6.4. Figure 6.2 compares the exact solution and the approximate solutions obtained for $n = 6$ and different values of $\alpha = 3.9, 3.95, 3.99$.

TABLE 6.3 The Absolute Errors $y(t)$ with $\alpha = 4$ and Different Values of n for Example 6.7

t	$n = 5$	$n = 8$
0	0	0
0.1	2.52028×10^{-4}	1.30488×10^{-6}
0.2	7.77547×10^{-4}	1.89061×10^{-6}
0.3	130262×10^{-3}	2.58505×10^{-7}
0.4	1.64352×10^{-3}	2.21313×10^{-6}
0.5	1.6996×10^{-3}	3.48887×10^{-6}
0.6	1.44678×10^{-3}	2.54364×10^{-6}
0.7	9.31915×10^{-4}	1.08668×10^{-7}
0.8	2.6831×10^{-4}	1.74064×10^{-6}
0.9	3.67471×10^{-4}	1.23787×10^{-6}
1	7.37686×10^{-4}	2.52194×10^{-7}

TABLE 6.4 The Absolute Errors $y(t)$ with $n = 6$ and Different Values of α for Example 6.7

t	$\alpha = 4$	$\alpha = 3.99$	$\alpha = 3.97$
0	7.0054×10^{-29}	1.797×10^{-28}	1.8537×10^{-28}
0.1	1.78422×10^{-7}	1.61818×10^{-7}	1.2537×10^{-7}
0.2	2.1029×10^{-6}	1.73183×10^{-6}	1.12669×10^{-6}
0.3	6.73797×10^{-6}	5.28321×10^{-6}	2.15744×10^{-6}
0.4	1.33202×10^{-5}	8.64241×10^{-6}	1.31277×10^{-5}
0.5	1.88213×10^{-5}	7.358×10^{-6}	1.68385×10^{-5}
0.6	1.99787×10^{-5}	3.56451×10^{-6}	5.29037×10^{-5}
0.7	1.47119×10^{-5}	2.7899×10^{-5}	1.16629×10^{-4}
0.8	2.78857×10^{-6}	6.71894×10^{-5}	2.12064×10^{-4}
0.9	1.45643×10^{-5}	1.20708×10^{-4}	3.39275×10^{-4}
1	3.57465×10^{-5}	1.86023×10^{-4}	4.93878×10^{-4}

6.4 GENOCCHI POLYNOMIALS

The Genocchi polynomials $G_n(t)$ of order n, are defined in interval $[0, 1]$, through the exponential generating functions [6]:

$$Q(r,t) = \frac{2re^{tr}}{e^r + 1} = \sum_{n=0}^{\infty} G_n(t)\frac{r^n}{n}, \quad |t| < \pi,$$

where $G_n(t)$ is given by

$$G_n(t) = \sum_{k=0}^{n} \binom{n}{k} g_{n-k} t^k, \tag{6.108}$$

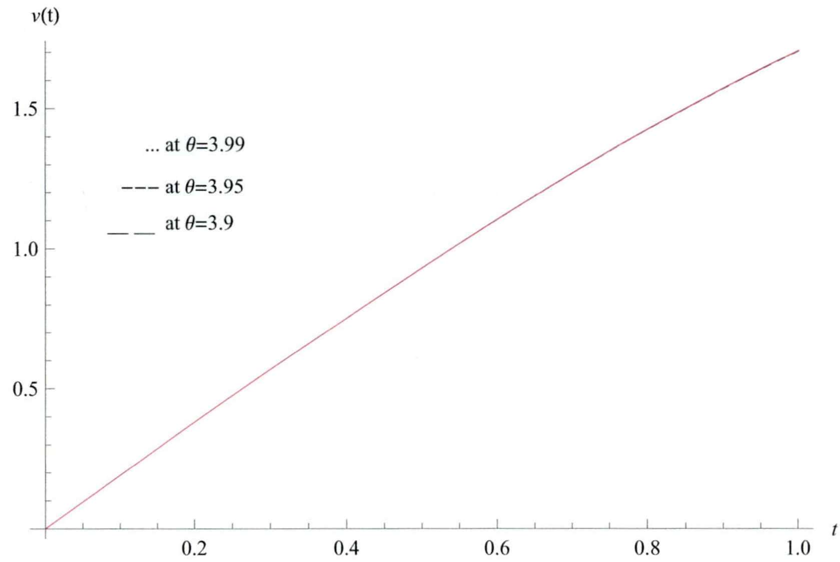

Figure 6.2 The exact solution (Red line) and approximation solutions of $y(t)$ when $n = 6$ for Example 6.7.

g_{n-k} is the Genocchi number which can also be obtained from

$$g_n = 2(1 - 2^n)B_n,$$

where B_n is the Bernoulli number.

Also, we have

$$\int_0^1 G_n(t)G_m(t)dt = \frac{2(-1)^n n!m!}{(m+n)!} g_{m+n}, \quad n, m \geqslant 1,$$

$$\frac{dG_n(t)}{dt} = nG_{n-1}(t), \quad n \geqslant 1,$$

We may approximate a function $y(t)$ in terms of Genocchi polynomials $G_n(t)$ as the basis. This can be done by the following infinite series:

$$y(t) = \sum_{i=0}^{\infty} c_i G_i(t),$$

In practice, we truncate the infinite series up to n number of Genocchi polynomials according to the desired accuracy of the problem, into the following truncated Genocchi series:

$$y(t) \simeq \sum_{i=0}^{n} c_i G_i(t) = C^T \Phi(t), \tag{6.109}$$

where $C = [c_1, c_2, \ldots, c_n]^T$ Genocchi coefficient matrix, and $\Phi(t) = [G_1(t), G_2(t), \ldots, G_n(t)]^T$ is the Genocchi basis matrix.

Thus, we can obtain

$$C = Q^{-1}\int_0^1 y(t)\Phi^T(t)dt, \tag{6.110}$$

where Q is an $n \times n$ matrix and is defined as

$$Q = \int_0^1 \Phi(t)\Phi^T(t)dt.$$

The vector $\Phi(t)$ can be written as

$$\Phi(t) = AT(t), \tag{6.111}$$

where $T(t) = [1, t, \ldots, t^n]$ and A is given in the following form:

$$A = \begin{bmatrix} \binom{1}{0}g_1 & \binom{1}{1}g_0 & 0 & \ldots & 0 \\ \binom{2}{0}g_2 & \binom{2}{1}g_1 & \binom{2}{2}g_0 & \ldots & 0 \\ \vdots & \vdots & \vdots & \ddots & \vdots \\ \binom{n}{0}g_n & \binom{n}{1}g_{n-1} & \binom{n}{2}g_{n-2} & \ldots & \binom{n}{n}g_0 \end{bmatrix}.$$

6.4.1 Operational Matrix of Fractional Integration

The fractional integral of $\Phi(t)$ can be approximated as [97]:

$$I^\alpha \Phi(t) \simeq P^\alpha \Phi(t), \tag{6.112}$$

where P^α is perational matrix of fractional integration based on Genocchi polynomials. Now, using Eq. (6.108), we calculate P^α as follows:

$$I^\alpha G_i(t) = \sum_{k=1}^{i} \frac{i!g_{i-k}}{(i-k)!k!} I^\alpha t^k = \sum_{k=1}^{i} \frac{i!g_{i-k}}{(i-k)!\Gamma(k+1+\alpha)} t^{k+\alpha}. \tag{6.113}$$

Now we approximate $t^{k+\alpha}$ for $k = 1, \ldots, n$, using terms of the Genocchi basis

$$t^{\alpha+k} = \sum_{j=1}^{n} \chi_{j,k} G_j(t). \tag{6.114}$$

Now, by substituting Eq. (6.114) in (6.113), we have

$$\begin{aligned} I^\alpha G_i(t) &= \sum_{j=1}^{N}\left(\sum_{k=1}^{i} \frac{i!g_{i-k}}{(i-k)!\Gamma(k+1+\alpha)}\chi_{j,k}\right) G_i(t) \\ &= \sum_{j=1}^{N}\left(\sum_{k=1}^{i} \theta_{i,j,k}\right) G_i(t), \end{aligned} \tag{6.115}$$

where

$$\theta_{i,j,k} = \frac{i!g_{i-k}}{(i-k)!\Gamma(k+1+\alpha)}\chi_{j,k}, \tag{6.116}$$

Equation (6.115) can be rewritten as

$$I^{\alpha}G_i(t) = \left[\sum_{k=1}^{i}\theta_{i,1,k}\ \sum_{k=1}^{i}\theta_{i,2,k}\ \ldots \sum_{k=1}^{i}\theta_{i,N,k}\right]\Phi(t), \qquad i=1,2,\ldots,n. \tag{6.117}$$

consequently, operational matrix of fractional integration can be obtained as:

$$P^{\alpha} = \begin{bmatrix} \theta_{1,1,1} & \theta_{1,2,1} & \ldots & \theta_{1,N,1} \\ \sum_{k=1}^{2}\theta_{2,1,k} & \sum_{k=1}^{2}\theta_{2,2,k} & \ldots & \sum_{k=1}^{2}\theta_{2,N,k} \\ \vdots & \vdots & \ldots & \vdots \\ \sum_{k=1}^{i}\theta_{i,1,k} & \sum_{k=1}^{i}\theta_{i,2,k} & \ldots & \sum_{k=1}^{i}\theta_{i,N,k} \\ \vdots & \vdots & \ldots & \vdots \\ \sum_{k=1}^{N}\theta_{N,1,k} & \sum_{k=1}^{N}\theta_{N,2,k} & \ldots & \sum_{k=1}^{N}\theta_{N,N,k} \end{bmatrix}. \tag{6.118}$$

6.4.2 Operational Matrix of Product

Let C is an arbitrary vector $n \times 1$, then $\hat{C}_{N\times N}$ is an operational matrix of the product as Ref. [97]

$$C^T\Phi(t)\Phi^T(t) \simeq \Phi^T(t)\hat{C}, \tag{6.119}$$

So, we have

$$\begin{aligned} C^T\Phi(t)\Phi^T(t) &= C^T\Phi(t)\left(T^T(t)A^T\right) \\ &= \left[C^T\Phi(t), t(C^T\Phi(t)), \ldots, t^{N-1}(C^T\Phi(t))\right]A^T \\ &= \left[\sum_{i=1}^{n} c_iG_i(t), \sum_{i=1}^{n} c_i(tG_i(t)), \ldots, \sum_{i=1}^{n} c_i(t^{N-1}G_i(t))\right]A^T. \end{aligned} \tag{6.120}$$

Now, we expand $t^kG_i(t)$ in terms of $\{G_i(t)\}_{i=1}^{n}$ for $i=1,2,\ldots,n, k=0,1,\ldots,n-1$.

$$t^kG_i(t) = \sum_{j=1}^{n} e_j^{k,i}G_j(t) = \mathbf{e}_{k,i}^T\Phi(t),$$

that $\mathbf{e}_{k,i} = \left[e_1^{k,i}, e_2^{k,i}, \ldots, e_n^{k,i}\right]^T$, then

$$\begin{aligned} \mathbf{e}_{k,i} &= Q^{-1}\left(\int_0^1 t^kG_i(t)\Phi(t)dt\right) \\ &= Q^{-1}\left[\int_0^1 t^kG_i(t)G_1(t)dt, \int_0^1 t^kG_i(t)G_2(t)dx, \ldots, \int_0^1 t^kG_i(t)G_n(t)dt\right]^T. \end{aligned}$$

Thus, we obtain finally

$$\begin{aligned}\sum_{i=1}^{n} c_i(t^k G_i(t)) &= \sum_{i=1}^{n} c_i \left(\sum_{j=1}^{n} e_j^{k,i} G_j(t)\right) = \sum_{j=1}^{n} G_j(t)\left(\sum_{i=1}^{n} c_i e_j^{k,i}\right) \\ &= \Phi(t)^T \left[\sum_{i=1}^{n} c_i e_1^{k,i}, \sum_{i=1}^{n} c_i e_2^{k,i}, \ldots, \sum_{i=1}^{n} c_i e_n^{k,i}\right]^T \\ &= \Phi^T(t)\left[\mathbf{e}_{k,1}, \mathbf{e}_{k,2}, \ldots, \mathbf{e}_{k,N}\right] C = \Phi^T(t)\widetilde{C_k}, \end{aligned} \tag{6.121}$$

where $\widetilde{C_k} = [\mathbf{e}_{k,1}, \mathbf{e}_{k,2}, \ldots, \mathbf{e}_{k,N}]\, C$, and $(k = 0, 1 \ldots, n-1)$.

If we define $\widetilde{C} = \left[\widetilde{C_1}, \widetilde{C_2}, \ldots, \widetilde{C_n}\right]$ then using Eqs. (6.120) and (6.121) we can write

$$\begin{aligned}C^T \Phi(t)\Phi^T(t) &= \left[\sum_{i=1}^{n} c_i G_i(t), \sum_{i=1}^{n} c_i(tG_i(t)), \ldots, \sum_{i=1}^{n} c_i(t^{n-1} G_i(t))\right] A^T \\ &= \Phi^T(t)\left[\widetilde{C_1}, \widetilde{C_2}, \ldots, \widetilde{C_n}\right] A^T = \Phi^T(t)\widetilde{C} A^T,\end{aligned}$$

so,

$$C^T \Phi(t)\Phi^T(t) \simeq \Phi^T(t)\widetilde{C} A^T.$$

On balance, the operational matrix of product is in the following form:

$$\hat{C} = \widetilde{C} A^T. \tag{6.122}$$

6.4.3 Applications

In this section, we use the Genocchi polynomials to solve the nonlinear fractional Abel differential equation of the first kind [97]:

$$D^\alpha y(t) = \kappa_3(t) y^3(t) + \kappa_2(t) y^2(t) + \kappa_1(t) y(t) + \kappa_0(t), \quad 0 < \alpha \le 1, \tag{6.123}$$

or the nonlinear fractional Abel equation of the second kind as

$$y(t) D^\alpha y(t) = \kappa_3(t) y^3(t) + \kappa_2(t) y^2(t) + \kappa_1(t) y(t) + \kappa_0(t), \quad 0 < \alpha \le 1, \tag{6.124}$$

where initial condition in this case:

$$y(0) = s_0, \tag{6.125}$$

where $\kappa_3(t) \neq 0, \kappa_2(t), \kappa_1(t)$ and $\kappa_0(t)$ are meromorphic functions.

At first, we assume that $D^\alpha y(t)$ is approximated by

$$D^\alpha y(t) \simeq Y^T \Phi(t). \tag{6.126}$$

By applying fractional integral on both sides of Eq. (6.126) and using Eqs. (6.118) and (6.125) $y(t)$ can be represented as

$$y(t) = I^\alpha D^\alpha y(t) + y(0) \simeq (Y^T P^\alpha + s^T)\Phi(t) = M_\alpha^T \Phi(t), \tag{6.127}$$

where $(Y^T P^\alpha + s^T) = M_\alpha^T$ and P^α is the operational matrix of integration of order α and

$$s^T = [y_0, 0, \ldots, 0].$$

Also, we approximate functions $\kappa_0(t), \kappa_1(t), \kappa_2(t), \kappa_3(t)$ by the Genocchi basis as

$$\kappa_0(t) \simeq K_0^T \Phi^T(t), \qquad \kappa_1(t) \simeq K_1^T \Phi^T(t), \tag{6.128}$$
$$\kappa_2(t) \simeq K_2^T \Phi^T(t), \qquad \kappa_3(t) \simeq K_3^T \Phi^T(t), \tag{6.129}$$

Now, by substituting equations (6.126)–(6.129) into Eq. (6.124) and using Eq. (6.122), we obtain

$$\begin{aligned} Y^T \Phi(t)\Phi^T(t) M_\alpha &= K_3^T \Phi(t)\Phi^T(t) \hat{M_\alpha}^2 M_\alpha + K_2^T \Phi(t)\Phi^T(t)\hat{M}_\alpha M_\alpha \\ &\quad + K_1^T \Phi(t)\Phi^T(t) M_\alpha + K_0^T \Phi(t). \end{aligned} \tag{6.130}$$

Also from Eq. (6.122), we have

$$Z^T \Phi(t)\Phi^T(t) \simeq \Phi^T(t)\hat{C}, \qquad K_3^T \Phi(t)\Phi^T(t) \simeq \Phi^T(t)\hat{K}_3, \tag{6.131}$$
$$K_2^T \Phi(t)\Phi^T(t) \simeq \Phi^T(t)\hat{K}_2, \qquad K_1^T \Phi(t)\Phi^T(t) \simeq \Phi^T(t)\hat{K}_1. \tag{6.132}$$

Now substituting Eqs. (6.131) and (6.132) in Eq. (6.130) we have

$$\Phi^T(t)\hat{C}M_\alpha = \Phi^T(t)\hat{K}_3 \hat{M_\alpha}^2 M_\alpha + \Phi^T(t)\hat{K}_2 \hat{M}_\alpha M_\alpha + \Phi^T(t)\hat{K}_1 M_\alpha + \Phi^T(t) K_0,$$

or

$$\Phi^T(t)\left(\hat{C}M_\alpha - \hat{K}_3 \hat{M_\alpha}^2 M_\alpha - \ddot{K}_2 \ddot{M}_\alpha M_\alpha - \dot{K}_1 M_\alpha - K_0\right) = 0.$$

Finally, we get

$$\left(\hat{C}M_\alpha - \hat{K}_3 \hat{M_\alpha}^2 M_\alpha - \hat{K}_2 \hat{M}_\alpha M_\alpha - \hat{K}_1 M_\alpha - K_0\right) = 0, \tag{6.133}$$

that by solving the above system, the vector C can be obtained. Consequently, $y(t)$ can be calculated from Eq. (6.127). Similarly, we can calculate $y(t)$ in Eq. (6.123).

Example 6.8 *Let us investigate the following fractional Abel differential equation of the second kind [97]*

$$y(t)D^\alpha y(t) + ty(t) + y^2(t) + t^2 y^3(t) = te^{-t} + t^2 e^{-3t}, \quad 0 < \alpha \leq 1, \tag{6.134}$$

with appropriate condition $y(0) = 1$. The accuracy solution is $y(t) = e^{-t}$.

By applying the technique described based on the Genocchi polynomials, The exact solution and the approximate solutions $y(t)$ for $n = 10$ and different values of $\alpha = 0.9, 0.97, 1$ are demonstated in Figure 6.3. Also the approximate solutions $y(t)$

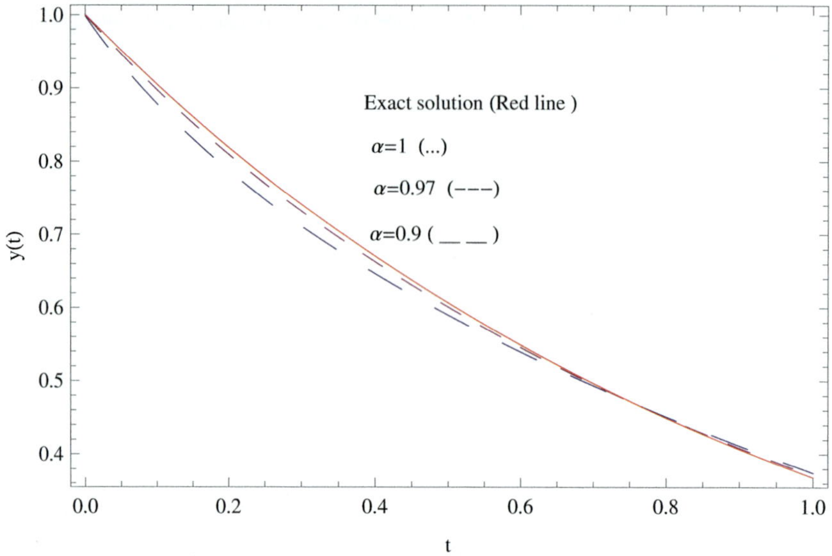

Figure 6.3 Exact and approximation solutions with $n = 10$ and $\alpha = 0.9, 0.97, 1$ for Example 6.8.

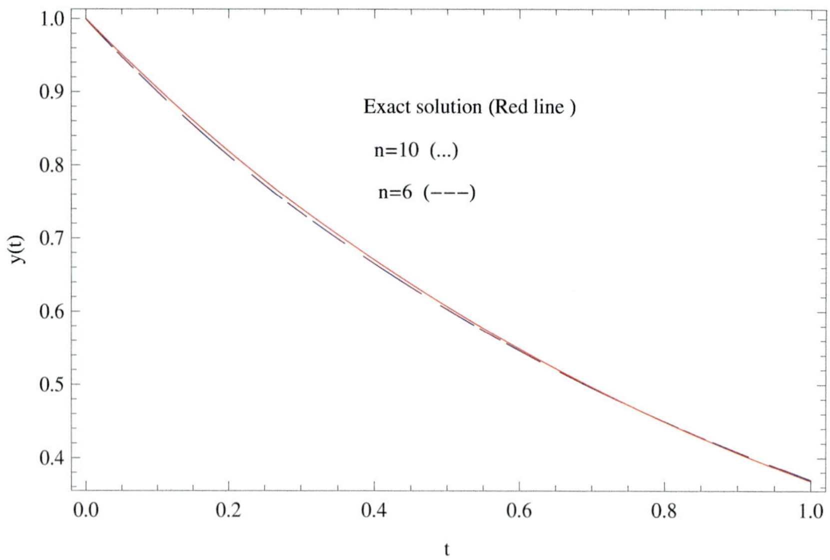

Figure 6.4 Exact and approximation solutions of $y(t)$ with $\alpha = 0.97$ and $n = 6, 10$ for Example 6.8.

together with the exact solution for $\alpha = 0.97$ and $n = 6, 10$ are plotted in Figure 6.4. Definitely, by increasing the value of n of Genocchi basis, the approximate values of $y(t)$ converge to the exact solutions. The absolute errors of the numerical results for $\alpha = 1$ and $n = 4, 6, 10$ are presented in Table 6.5.

TABLE 6.5 The Absolute Errors of $y(t)$ with $\alpha = 1$ and Different Values of n for Example 6.8

t	$n = 10$	$n = 6$	$n = 4$
0	4.22524×10^{-6}	-5.31722×10^{-6}	-6.45786×10^{-3}
0.1	4.54361×10^{-7}	2.23431×10^{-6}	7.08418×10^{-5}
0.2	3.67411×10^{-7}	8.18054×10^{-7}	2.56526×10^{-4}
0.3	1.18684×10^{-6}	1.43944×10^{-6}	1.80473×10^{-4}
0.4	5.99053×10^{-7}	1.8424×10^{-6}	2.56526×10^{-5}
0.5	9.44672×10^{-7}	5.79925×10^{-7}	8.22628×10^{-3}
0.6	9.20404×10^{-7}	1.07731×10^{-6}	1.10107×10^{-4}
0.7	6.04158×10^{-7}	1.90827×10^{-6}	6.85608×10^{-5}
0.8	1.36001×10^{-6}	1.26478×10^{-6}	2.75025×10^{-5}
0.9	9.00125×10^{-7}	9.55465×10^{-7}	1.06553×10^{-4}
1	4.20447×10^{-6}	4.83369×10^{-6}	4.70342×10^{-4}

CHAPTER 7

Wavelet Method for Fractional Differential Equations

WAVELET ANALYSIS is a new branch of mathematics is widely applied in signal analysis, image processing, numerical analysis, etc. The name wavelets means small waves (the sinusoids used in Fourier analysis are 'big waves'), and, in short, a wavelet is an oscillation that decays quickly. An excellent account of the history is in Ref. [35]. Wavelets are a powerful new mathematical tool, which offers the possibility to treat in a natural way quantities characterized by several length scales.

Today, wavelets are not only the workspace in computer imaging and animation, but they are also used by the FBI to encode its data base of million fingerprints. In future, scientist may put wavelet analysis for diagnosing breast cancer, looking for heart abnormalities, predicting the weather, signal processing, data compression, smoothing and image compression, fingerprints verification, DNA analysis, protein analysis, blood pressure, heart rate and ECG analysis, finance, internet traffic description, speech recognition and computer graphics [80].

Wavelets constitute a family of functions constructed from the dilation and translation of a single function called the mother wavelet. When the dilation parameter a and the translation parameter b vary continuously, we have the following family of continuous wavelets as, Refs. [46, 74]:

$$\psi_{a,b}(t) = |a|^{-\frac{1}{2}} \psi\left(\frac{t-b}{a}\right), \qquad a, b \in R, \ a \neq 0. \tag{7.1}$$

If we restrict the parameters a and b to discrete values as $a = a_0^k$, $b = nb_0a_0^k$, $a_0 > 1$, $b_0 > 0$ and $n, \text{and} \, k$ positive integers, we have the following family of discrete wavelets:

$$\psi_{k,n}(t) = |a_0|^{\frac{k}{2}} \psi\left(a_0^k t - nb_0\right),$$

where $\psi_{k,n}(t)$ form a wavelet basis for $L^2(R)$. In particular, when $a_0 = 2$ and $b_0 = 1$ then $\psi_{k,n}(t)$ forms an orthonormal basis [46]. The larger the number k, the finer the resolution or wavelets.

DOI: 10.1201/9781003474609-7

7.1 HAAR WAVELET

Among the wavelet families, which are defined by an analytical expression, special attention deserves the Haar wavelets. In 1910, Alfred Haar introduced the notion of wavelets. His initial theory has been expanded recently into a wide variety of applications, but primarily it allows for the representation of various functions by a combination of step functions and wavelets over specified interval widths.

Haar wavelets are mathematically the simplest among all the wavelet families. Due to the simplicity, the Haar wavelets are very effective for solving fractional ordinary differential and fractional partial differential equations. Therefore the idea, to apply Haar wavelet technique also for solving the fractional differential equation, arises. The method with far less degrees of freedom and with smaller CPU time provides better solutions than classical ones. Haar wavelet method (HWM), for solving the fractional differential, integral and integro-differential equations, exhibit several advantageous features

i. Very high accuracy fast transformation and the possibility of implementation of fast algorithms compared with other known methods.

ii. The simplicity and small computation costs, resulting from the sparsity of the transform matrices and the small number of significant wavelet coefficients.

iii. The method is also very convenient for solving the boundary value problems, since the boundary conditions are taken care of automatically.

In practical applications, a small number of terms increase the calculation speed and saves memory storage; a large number of terms improve resolution accuracy. Therefore, a trade-off between calculation speed, memory saving and the resolution accuracy has been considered in the analysis. For larger m (that is $m = 16$, $m = 32$, $m = 64$, $m = 128$, $m = 256$), the numerical solution is more and more close to the exact solution. The Haar results show that the Haar wavelet method is quite reasonable when compared to exact solution. These calculations demonstrate that the accuracy of this method is quite high.

Haar wavelets is the oldest and most basic of the wavelet systems. The orthogonal set of Haar wavelets is a group of square waves with magnitudes ± 1 some intervals and zero elsewhere [74, 114]

$$\psi(t) = \begin{cases} 1, & \text{for } [0, \frac{1}{2}), \\ -1, & \text{for } [\frac{1}{2}, 1), \\ 0, & \text{elsewhere.} \end{cases} \tag{7.2}$$

To perform wavelet transform, Haar wavelet uses dilations and translations of function, that is, the transform makes the following function.

$$h_n = \psi(2^j t - k) \quad for \quad n \geq 1,$$

we have $h_0(t) = 1$, with $0 \leq t < 1$ and zero otherwise, i.e., the normalized scaling function, $n = 2^j + k$ for $j \geq 0$ and $0 \leq k \leq 2^j - 1$.

Any function $y(t)$ which is square integrable in the interval $[0,1)$ can be expanded in a Haar series

$$y(t) = \sum_{i=0}^{\infty} c_i h_i(t), \tag{7.3}$$

where the Haar coefficients $c_i, i = 0, 1, 2, \ldots,$ are determined as follows:

$$c_i = 2^j \int_0^1 y(t) h_i(t) dt, \tag{7.4}$$

by applying the orthogonal property of Haar wavelet

$$\int_0^1 h_l(t) h_i(t) dt = \begin{cases} 2^{-j}, & l = i, \\ 0, & l \neq i, \end{cases} \tag{7.5}$$

Since it is not realistic to use an infinite number of wavelets to represent the function $f(t)$, Eq. (7.3) terminates at a finite term:

$$y(t) \simeq \sum_{i=0}^{m-1} c_i h_i(t) = C^T H_m(t) = \hat{y}(t), \tag{7.6}$$

where $m = 2^{j+1}$. $\hat{y}(t)$ denotes the truncated sum. The Haar coefficient vector C and Haar function vector $H(t)$ are defined as

$$\begin{aligned} C &= [c_0, c_1, \ldots, c_{m-1}]^T, \\ H_m(t) &= [h_0(t), h_1(t), \ldots, h_{m-1}(t)]^T. \end{aligned} \tag{7.7}$$

Taking the collocation points as follows:

$$t_l = \frac{(2l-1)}{2m}, \quad (l = 1, 2, \ldots, m), \tag{7.8}$$

we defined the m-square Haar matrix $\Phi_{m\times m}$ as:

$$\Phi_{m\times m} = \left[H_m\left(\frac{1}{2m}\right) \; H_m\left(\frac{3}{2m}\right) \; \cdots \; H_m\left(\frac{2m-1}{2m}\right) \right]. \tag{7.9}$$

Correspondingly, we have

$$\hat{y}_m = \left[\hat{y}\left(\frac{1}{2m}\right) \; \hat{y}\left(\frac{3}{2m}\right) \; \cdots \; \hat{y}\left(\frac{2m-1}{2m}\right) \right] = C^T \Phi_{m\times m}, \tag{7.10}$$

as the m-square Haar matrix $\Phi_{m\times m}$ is an invertible matrix, the Haar coefficient vector C^T can be obtained by $C^T = \hat{y}_m \Phi_{m\times m}^{-1}$.

7.1.1 Operational Matrix of Fractional Integration

The integration of the $H_m(t)$ defined in Eq. (7.7) can be approximated by Haar series with Haar coefficient matrix P [46].

$$\int_0^t H_m(\tau)d\tau \simeq P_{m\times m}H_m(t), \tag{7.11}$$

where the m-square matrix P is called Haar wavelet operational matrix of integration. A subscript $m \times m$ of P denotes its dimension

$$P_{m\times m} = \frac{1}{2m}\begin{bmatrix} 2mP_{m/2\times m/2} & -H_{m/2\times m/2} \\ H^{-1}_{m/2\times m/2} & 0_{m/2\times m/2} \end{bmatrix}, \qquad P_{1\times 1} = \frac{1}{2}. \tag{7.12}$$

Next, we generalize the Haar wavelet operational matrix to fractional integration. If $y(t)$ is expanded by Haar series, as shown in Eq. (7.6), the Riemann-Liouville fractional integration can rewrite

$$I^\alpha y(t) \simeq \frac{C_m^T}{\Gamma(\alpha)}\int_0^t (t-\tau)^{\alpha-1}H_m(\tau)d\tau, \tag{7.13}$$

Also, we introduce a m-set of block pulse functions (BPF) defined by:

$$b_i(t) = \begin{cases} 1, & \frac{i}{m} \le t < \frac{(i+1)}{m}, \\ 0, & \text{otherwise}, \end{cases} \tag{7.14}$$

where $i = 0, 1, 2, \ldots, (m-1)$.

The functions $b_i(t)$ are disjoint and orthogonal. That is,

$$b_i(t)b_l(t) = \begin{cases} 0, & i \neq l, \\ b_i(t), & i = l. \end{cases} \tag{7.15}$$

$$\int_0^1 b_i(\tau)b_l(\tau) = \begin{cases} 0, & i \neq l, \\ \frac{1}{m}, & i = l. \end{cases} \tag{7.16}$$

As the Haar functions are piecewise constant, it may be expanded into an m-term BPF as

$$H_m(t) = \Phi_{m\times m}B_m(t), \tag{7.17}$$

where $B_m(t) = [b_0(t)\ b_1(t)\ \ldots\ b_i(t)\ \ldots b_{m-1}(t)]^T$.

Kilicman and Zhour have given Block Pulse operational matrix of the fractional order integration F^α as Ref. [70] following

$$I^\alpha B_m(t) \simeq F^\alpha B_m(t). \tag{7.18}$$

where

$$F^\alpha = \frac{1}{m^\alpha}\frac{1}{\Gamma(\alpha+2)}\begin{bmatrix} 1 & \xi_1 & \xi_2 & \dots & \xi_{m-1} \\ 0 & 1 & \xi_1 & \dots & \xi_{m-2} \\ 0 & 0 & 1 & \dots & \xi_{m-3} \\ 0 & 0 & 0 & \ddots & \vdots \\ 0 & 0 & 0 & 0 & 1 \end{bmatrix}, \tag{7.19}$$

with $\xi_k = (k+1)^{\alpha+1} - 2k^{\alpha+1} + (k-1)^{\alpha+1}$.

Here, we introduce as follows:

Let

$$I^\alpha H_m(t) \simeq P^\alpha_{m\times m} H_m(t), \tag{7.20}$$

where the m-square matrix $P^\alpha_{m\times m}$ is called Haar wavelet operational matrix of the fractional integration.

Using Eqs. (7.17) and (7.20), we have

$$I^\alpha H_m(t) \simeq I^\alpha \Phi_{m\times m} B_m(t) = \Phi_{m\times m} I^\alpha B_m(t) \simeq \Phi_{m\times m} F^\alpha B_m(t). \tag{7.21}$$

From Eqs. (7.20) and (7.21) we get

$$P^\alpha_{m\times m} H_m(t) = P^\alpha_{m\times m}\Phi_{m\times m} B_m(t) = \Phi_{m\times m} F^\alpha B_m(t), \tag{7.22}$$

then, the Haar wavelet operational matrix of the fractional integration $P^\alpha_{m\times m}$ is given by

$$P^\alpha_{m\times m} = \Phi_{m\times m} F^\alpha \Phi^{-1}_{m\times m}. \tag{7.23}$$

7.1.2 Applications

Example 7.1 *Let us consider the Bagley-Torvik [74]*

$$D^2 y(t) + D^{1.5} y(t) + y(t) = f(t), \quad t > 0. \tag{7.24}$$

Subject to

$$y(0) = y'(0) = 1, \tag{7.25}$$

where $f(t) = 1 + t$ and the exact solution of this problem is $y(t) = 1 + t$.

Let

$$D^2 y(t) = C^T H_m, \tag{7.26}$$

together with the initial states, then we have

$$D^{1.5} y(t) = I^{0.5} D^2 y(t) = C^T P^{0.5}_{m\times m} H_m(t) = C^T \Phi_{m\times m} F^{0.5} \Phi^{-1}_{m\times m} H_m(t), \tag{7.27}$$

$$Dy(t) = C^T P^1_{m\times m} H_m(t) + y'(0) = C^T P^1_{m\times m} H_m(t) + 1, \tag{7.28}$$

$$y(t) = C^T P^2_{m\times m} H_m(t) + [1, 1, \dots, 1]\Phi^{-1}_{m\times m} P^1_{m\times m} H^T_m + y(0). \tag{7.29}$$

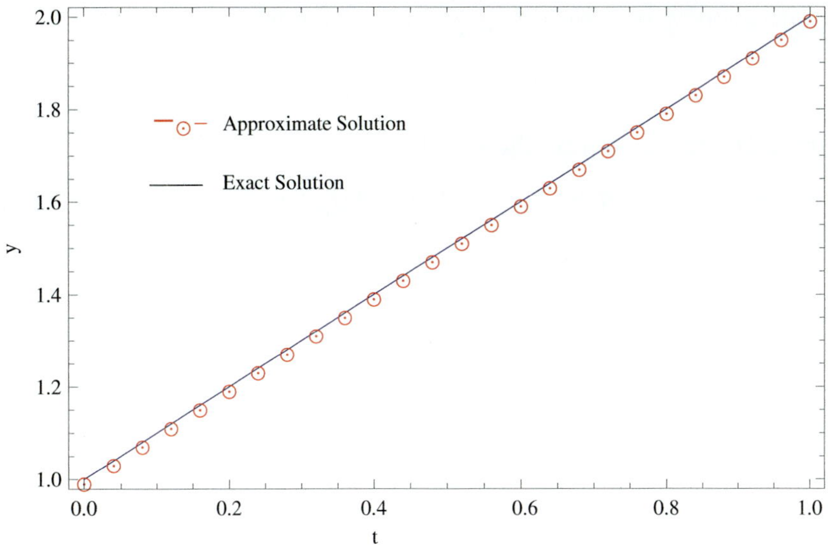

Figure 7.1 Numerical and exact solution with $m = 32$ for Example 7.1.

Similarly, the input signal $f(t)$ may be expanded by the Haar functions as follows:

$$f(t) = f^T H_m(t), \tag{7.30}$$

where f^T is a known constant vector. Substituting Eqs. (7.26)–(7.30) into Eq. (7.24), we get

$$C^T H_m(t) + C^T P^{0.5}_{m\times m} H_m(t) + \Big(C^T P^2_{m\times m} H_m(t)$$

$$+[1, 1, \ldots, 1]\Phi^{-1}_{m\times m} P^1_{m\times m} H_m(t) + [1, 1, \ldots, 1] \Big) = f^T H_m(t). \tag{7.31}$$

By solving the system of algebraic equations (7.31), we can obtain the coefficients C^T. Then using Eq. (7.29), we can get the output response $y(t)$. The numerical result for $m = 32$ is shown in Figure 7.1. The numerical solution is in perfect agreement with the exact solutions.

Example 7.2 *Consider the fractional order variable coefficient linear differential equation [74]*

$$aD^2 y(t) + b(t)D^{\alpha_2} y(t) + c(t)Dy(t) + e(t)D^{\alpha_1} y(t) + k(t)y(t) = f(t), \tag{7.32}$$

where $0 < \alpha_1 \le 1,\ 1 < \alpha_2 \le 2$, *and*

$$f(t) = a - \frac{b(t)}{\Gamma(3-\alpha_2)} t^{2-\alpha_2} - c(t)t - \frac{e(t)}{\Gamma(3-\alpha_1)} t^{2-\alpha_1} + k(t)\left(2 - \frac{1}{2}t^2\right), \tag{7.33}$$

subject to

$$y(0) = 2, \quad y'(0) = 0, \tag{7.34}$$

the exact solution of this problem is $y(t) = 2 - \frac{1}{2}t^2$.

We approximate $D^2y(t)$ by the Haar functions as follows:

$$D^2y(t) = R^T H_m(t), \tag{7.35}$$

together with the initial states, then we have

$$D^{\alpha_2}y(t) = R^T P_{m\times m}^{2-\alpha_2} H_m(t), \tag{7.36}$$
$$Dy(t) = R^T P_{m\times m}^{1} H_m(t), \tag{7.37}$$
$$D^{\alpha_1}y(t) = R^T P_{m\times m}^{2-\alpha_1} H_m(t), \tag{7.38}$$
$$y(t) = R^T P_{m\times m}^{2} H_m(t) + 2. \tag{7.39}$$

Similarly, $f(t)$ may be expanded by the Haar functions as follows:

$$f(t) = f^T H_m(t), \tag{7.40}$$

where f^T is a known constant vector. Substituting Eqs. (7.35)–(7.40) into Eq. (7.32), we have

$$\begin{aligned} &aR_m^T H_m(t) + b(t)R_m^T P_{m\times m}^{2-\alpha_2} H_m(t) + c(t)R_m^T P^1 m \times m H_m(t) \\ &+ e(t)R_m^T P_{m\times m}^{2-\alpha_2} H_m(t) + k(t)\left[R_m^T P_{m\times m}^{2} H_m(t) + 2\right] = f_m^T H_m(t). \end{aligned} \tag{7.41}$$

There are different methods for processing these variable coefficients of Eq. (7.41). Here we apply the collocation method. Taking the collocation points as follows:

$$t_i = \frac{(2i-1)}{2m}, \quad i = 1, 2, \ldots, m, \tag{7.42}$$

together with Eq. (7.17), we can transform Eq. (7.32) into a system of algebraic equations. By solving the system of algebraic equations, we can obtain the coefficients R_m^T. Then using Eq. (7.39), we can get the output response $y(t)$. For

$$a = 1, b(t) = t^{\frac{1}{2}}, c(t) = t^{\frac{1}{3}}, e(t) = t^{\frac{1}{4}}, k(t) = t^{\frac{1}{5}}, \alpha_1 = 0.333, \alpha_2 = 1.234,$$

we present numerical values of the solution to Eq. (7.32) by Haar wavelet method with $m = 64$ and exact solution in Figure 7.2, which is in perfect agreement with the exact solutions.

Example 7.3 *We consider the nonlinear fractional order differential equation [74]*

$$aD^{2.2}y(t) + bD^{\alpha_2}y(t) + cD^{\alpha_1}y(t) + e[y(t)]^3 = f(t) \quad 0 < \alpha_1 \le 1, \quad 1 < \alpha_2 \le 2, \tag{7.43}$$

and

$$f(t) = \frac{2a}{\Gamma(1.8)}t^{0.8} + \frac{2b}{\Gamma(4-\alpha_2)}t^{3-\alpha_2} + \frac{2c}{\Gamma(4-\alpha_1)}t^{3-\alpha_1} + e\left[\frac{1}{3}t^3\right]^3, \tag{7.44}$$

subject to

$$y(0) = y'(0) = y''(0) = 0. \tag{7.45}$$

The exact solution of this problem is $y(t) = \frac{1}{3}t^3$.

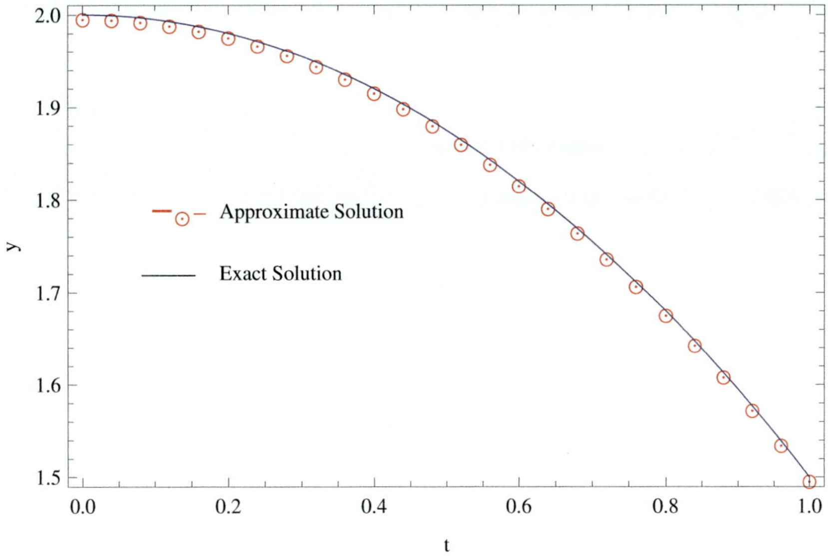

Figure 7.2 Numerical and exact solution with $m = 64$ for Example 7.2.

We approximate $D^{2.2}y(t)$ by the Haar functions as follows:

$$D^{2.2}y(t) = K^T H_m(t), \tag{7.46}$$

together with the initial states, then we have

$$D^{\alpha_2}y(t) = K^T P^{2.2-\alpha_2}_{m\times m} H_m(t), \tag{7.47}$$
$$D^{\alpha_1}y(t) = K^T P^{2.2-\alpha_1}_{m\times m} H_m(t), \tag{7.48}$$
$$y(t) = K^T P^{2.2}_{m\times m} H_m(t). \tag{7.49}$$

Since $H_m(t) = \Phi_{m\times m} B_m(t)$, from Eq. (7.49) we have

$$y(t) = K^T P^{2.2}_{m\times m} \Phi_{m\times m} B_m(t). \tag{7.50}$$

Let

$$K^T P^{2.2}_{m\times m} \Phi_{m\times m} = [a_1, a_2, \ldots, a_m], \tag{7.51}$$

then

$$[y(t)]^3 = \left[a_1^2, a_2^3, \ldots, a_m^3\right] B_m(t). \tag{7.52}$$

Similarly, the input signal $f(t)$ may be expanded by the Haar functions as follows:

$$f(t) = f^T H_m(t), \tag{7.53}$$

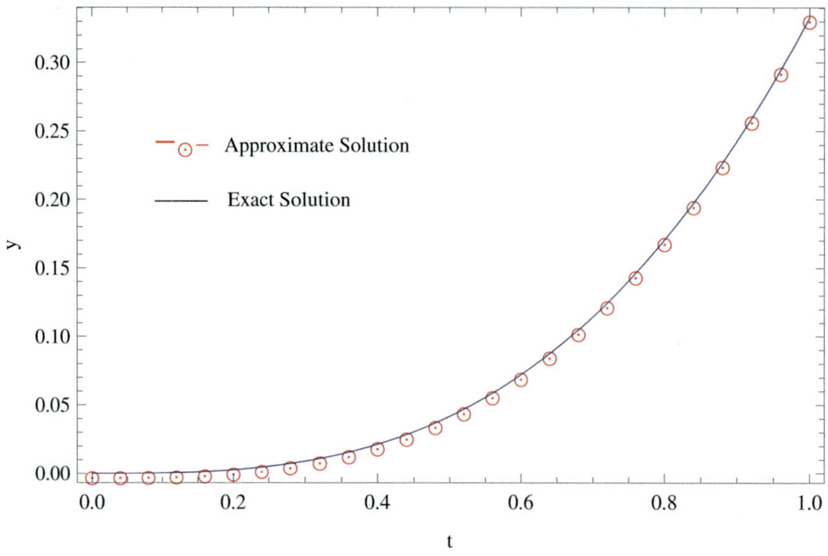

Figure 7.3 Numerical and exact solution with $m = 16$ for Example 7.3.

where f^T is a known constant vector. Substituting Eqs. (7.46)–(7.48) and (7.52) into Eq. (7.43), together with $H_m(t) = \Phi_{m\times m}B_m(t)$ we have

$$\begin{aligned}&K^T\Phi_{m\times m}B_m(t) + K^T P^{2.2-\alpha_2}_{m\times m}\Phi_{m\times m}B_m(t)B_m(t)\\&+K^T P^{2.2-\alpha_1} m \times m\Phi_{m\times m}B_m(t)B_m(t)\left[a_1^3, a_2^3, \ldots, a_m^3\right]B_m(t)\\&-f^T\Phi_{m\times m}B_m(t) = 0.\end{aligned} \tag{7.54}$$

For $a = 1, b = 1, c = 1, e = 1, \alpha_1 = 0.75, \alpha = 1.25$, Figure 7.3 shows a behavior of the numerical solution for $m = 16$, which is in agreement with the exact solutions.

7.2 CHEBYSHEV WAVELET

Chebyshev wavelets $\psi(t)$, on the interval $[0, 1)$ are defined as Refs. [37, 73, 107]

$$\psi_{n,m}(t) = \begin{cases} 2^{k/2}\tilde{T}_m(2^k t - 2n + 1), & \frac{n-1}{2^{k-1}} \le t < \frac{n}{2^{k-1}}, \\ 0, & \text{otherwise}, \end{cases} \tag{7.55}$$

where

$$\tilde{T}_m = \begin{cases} \frac{1}{\sqrt{\pi}}, & m = 0, \\ \sqrt{\frac{2}{\pi}}T_m(t), & \text{otherwise}, \end{cases} \tag{7.56}$$

and $m = 0, 1, \ldots, M-1, n = 1, 2, \ldots, 2^{k-1}, k$ is any positive integer and $T_m(t)$ are Chebyshev polynomials of the first kind of degree m which are orthogonal with respect to the weight function $\omega(t) = \frac{1}{\sqrt{1-t^2}}$, on the interval $[-1, 1]$ and $T_m(t)$ can be determined by the following recurrence formula:

$$T_0(t) = 1, \quad T_1(t) = t, \quad T_{m+1}(t) = 2tT_m(t) - T_{m-1}(t), \quad m = 1, 2, 3, \ldots. \tag{7.57}$$

It was noticed that Chebyshev wavelets are orthogonal concerning the weight function $\omega_n(t) = \omega(2^{k+1}t - 2n + 1)$.

A function $y(t)$ defined over $[0, 1)$ may be expanded as

$$y(t) = \sum_{n=1}^{\infty} \sum_{m=0}^{\infty} c_{nm}\psi_{nm}(t), \tag{7.58}$$

where $c_{nm} =< y(t), \psi_{nm}(t) >$, in which $<,>$ denotes the inner product. If the infinite series in Eq. (7.58) is truncated, then Eq. (7.58) can be written as

$$y(t) \simeq \sum_{n=1}^{2^{k-1}} \sum_{m=0}^{M-1} c_{nm}\psi_{nm}(t) = C^T\Psi(t), \tag{7.59}$$

where T indicates transposition, C and $\Psi(t)$ are $2^{k-1}M \times 1$ matrices given by

$$C = \left[c_{1,0}, c_{1,1}, \ldots, c_{1,M-1}, c_{2,0}, \ldots, c_{2,M-1}, \ldots, c_{2^{k-1},0}, \ldots, c_{2^{k-1},M-1}\right]^T. \tag{7.60}$$

$$\Psi(t) = \left[\psi_{1,0}, \psi_{1,1}, \ldots, \psi_{1,M-1}, \psi_{2,0}, \ldots, \psi_{2,M-1}, \psi_{2^{k-1},0}, \ldots, \psi_{2^{k-1},M-1}\right]^T. \tag{7.61}$$

Taking the collocation points as follows:

$$t_i = \frac{(2i-1)}{2^k M}, \quad i = 1, 2, \ldots, 2^{k-1}M. \tag{7.62}$$

We define the Chebyshev wavelet matrix $\Phi_{m \times m}$ as:

$$\Phi_{m \times m} = [\Psi(t_1), \Psi(t_2), \ldots, \Psi(t_m)], \tag{7.63}$$

where $m = 2^{k-1}M$.

7.2.1 Operational Matrix of Fractional Integration

The integration of the vector $\Psi(t)$ defined in Eq. (7.61) can be obtained as

$$\int_0^t \Psi(\tau)d\tau \simeq P\Psi(t), \tag{7.64}$$

where P is the $2^{k-1}M \times 2^{k-1}M$ operational matrix for integration [107].

The fractional integration of order α of vector $\Psi(t)$ in Eq. (7.61) can be shown as

$$I^\alpha \Psi(t) \approx P^\alpha_{m \times m}\Psi(t), \tag{7.65}$$

where matrix $P^\alpha_{m \times m}$ is called Chebyshev wavelet operational matrix of the fractional integration [73]. For obtaining, we define a m-set of block pulse functions (BPF) as:

$$b_i(t) = \begin{cases} 1, & \frac{i}{m} \leq t < \frac{(i+1)}{m}, \\ 0, & \text{otherwise}, \end{cases} \tag{7.66}$$

where $i = 0, 1, 2, \ldots, (m-1)$.

The functions $b_i(t)$ are disjoint and orthogonal. That is,

$$b_i(t)b_l(t) = \begin{cases} 0, & i \neq l, \\ b_i(t), & i = l. \end{cases} \tag{7.67}$$

$$\int_0^1 b_i(\tau)b_l(\tau) = \begin{cases} 0, & i \neq l, \\ \frac{1}{m}, & i = l. \end{cases} \tag{7.68}$$

Similarly, Chebyshev wavelets may be expanded into an m-term block pulse functions (BPF) as

$$\Psi(t) = \Phi_{m\times m}B_m(t), \tag{7.69}$$

where $B_m(t) = [b_0(t)\ b_1(t)\ \ldots\ b_i(t)\ \ldots b_{m-1}(t)]^T$.

In Ref. [70], Kilicman and Zhour have given Block Pulse operational matrix of the fractional order integration F^α as follows:

$$I^\alpha B_m(t) \approx F^\alpha B_m(t), \tag{7.70}$$

where

$$F^\alpha = \frac{1}{m^\alpha}\frac{1}{\Gamma(\alpha+2)}\begin{bmatrix} 1 & \xi_1 & \xi_2 & \cdots & \xi_{m-1} \\ 0 & 1 & \xi_1 & \cdots & \xi_{m-2} \\ 0 & 0 & 1 & \cdots & \xi_{m-3} \\ 0 & 0 & 0 & \ddots & \vdots \\ 0 & 0 & 0 & 0 & 1 \end{bmatrix}, \tag{7.71}$$

with $\xi_k = (k+1)^{\alpha+1} - 2k^{\alpha+1} + (k-1)^{\alpha+1}$.

Using Eqs. (7.69) and (7.70), we have

$$I^\alpha\Psi(t) \approx I^\alpha\Phi_{m\times m}B_m(t) = \Phi_{m\times m}I^\alpha B_m(t) \approx \Phi_{m\times m}F^\alpha B_{m\times m}(t). \tag{7.72}$$

From Eqs. (7.65) and (7.72) we get

$$P^\alpha_{m\times m}\Psi(t) = P^\alpha_{m\times m}\Phi_{m\times m}I^\alpha B_m(t) = \Phi_{m\times m}F^\alpha B_{m\times m}(t). \tag{7.73}$$

Then, the Chebyshev wavelet operational matrix of the fractional integration $P^\alpha_{m\times m}$ is given by

$$P^\alpha_{m\times m} = \Phi_{m\times m}F^\alpha\Phi^{-1}_{m\times m}. \tag{7.74}$$

7.2.2 Applications

Example 7.4 *We consider fractional Riccati equation [73]*

$$D^\alpha y(t) = 2y(t) - y^2(t) + 1, \quad 0 < \alpha \leq 1, \quad 0 \leq t < 5, \tag{7.75}$$

subject to the initial state $y(t) = 0$. The exact solution of this problem, when $\alpha = 1$, is

$$y(t) = 1 + \sqrt{2}\tanh\left(\sqrt{2}t + \frac{1}{2}\ln\left(\frac{\sqrt{2}-1}{\sqrt{2}+1}\right)\right), \tag{7.76}$$

Here we use the Chebyshev wavelet operational matrices of the fractional integration to solve it. Let

$$D^\alpha y(t) = K^T \Psi_m(t), \tag{7.77}$$

together with the initial states, then we have

$$y(t) = K^T P^\alpha_{m\times m} \Psi_m(t). \tag{7.78}$$

Since $\Psi_m(t) = \Phi_{m\times m} B_m(t)$, from Eq. (7.78) we have

$$y(t) = K^T P^\alpha_{m\times m} \Phi_{m\times m} B_m(t). \tag{7.79}$$

Let

$$K^T P^\alpha_{m\times m} \Phi_{m\times m} = [a_1, a_2, \ldots, a_m], \tag{7.80}$$

and using Eq. (7.66), we have

$$[y(t)]^2 = [a_1 b_1(t) + a_2 b_2(t) + \ldots + a_m b_m(t)]^2 = \left[a_1^2, a_2^2, \ldots, a_m^2\right] B_m(t). \tag{7.81}$$

Substituting Eqs. (7.77), (7.78) and (7.81) into Eq. (7.75), we have

$$\begin{aligned} &K^T P^\alpha_{m\times m} \Phi_{m\times m} B_m(t) + \left[a_1^2, a_2^2, \ldots, a_m^2\right] B_m(t) \\ &\quad - 2K^T P^\alpha_{m\times m} \Phi_{m\times m} B_m(t) - [1, 1, \ldots, 1] B_m(t) = 0. \end{aligned} \tag{7.82}$$

The numerical solution, for $m = 96$, is shown in Figure 7.4. The numerical solution for $\alpha = 1$ and $m = 192$ is presented in Table 7.1.

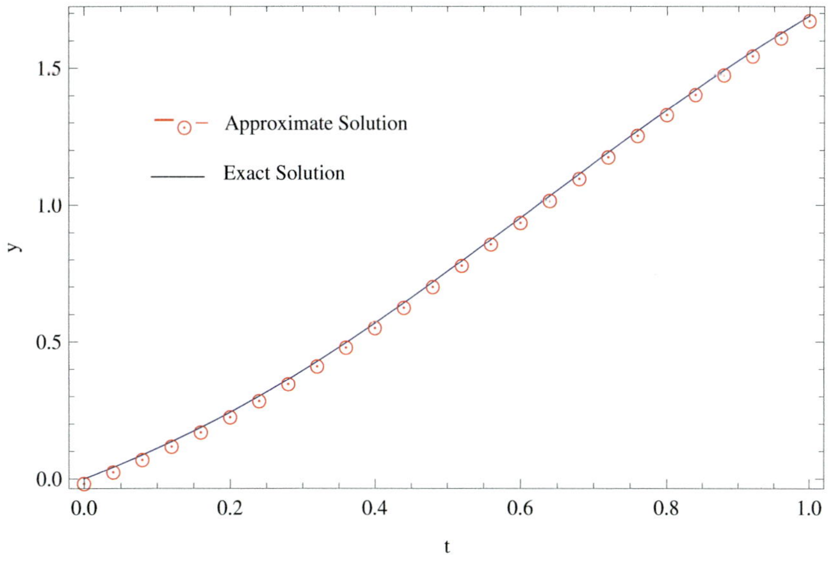

Figure 7.4 Numerical solution and exact solution of $y(t)$ for $m = 96$ and $\alpha = 1$ for Example 7.4.

TABLE 7.1 Numerical Results of $y(t)$ for $\alpha = 1$ and $m = 192$ for Example 7.4

t	$\alpha = 1$	Exact Solution
0.1	0.11.311	0.110295
0.2	0.241995	0.241977
0.3	0.395123	0.395105
0.4	0.567829	0.567812
0.5	0.756029	0.756014
0.6	0.953576	0.953566
0.7	1.152955	1.152949
0.8	1.346365	1.346364
0.9	1.526909	1.526911
1	1.689494	1.689498

Example 7.5 *we consider the nonlinear fractional order differential equation [73]*

$$aD^2y(t) + bD^{\alpha_2}y(t) + cD^{\alpha_1}y(t) + e[y(t)]^3 = f(t), \quad 0 < \alpha_1 \leq 1, \quad 1 < \alpha_2 \leq 2, \tag{7.83}$$

and

$$f(t) = \frac{2a}{\Gamma(1.8)}t^{0.8} + \frac{2b}{\Gamma(4-\alpha_2)}t^{3-\alpha_2} + \frac{2c}{\Gamma(4-\alpha_1)}t^{3-\alpha_1} + e\left[\frac{1}{3}t^3\right]^3, \tag{7.84}$$

subject to

$$y(0) = y'(0) = 0. \tag{7.85}$$

The exact solution of this problem is $y(t) = \frac{1}{3}t^3$.

Let

$$D^2y(t) = K^T\Psi_m(t), \tag{7.86}$$

together with the initial states, then we have

$$D^{\alpha_2}y(t) = K^T P_{m\times m}^{2-\alpha_2}\Psi_m(t), \tag{7.87}$$

$$D^{\alpha_1}y(t) = K^T P_{m\times m}^{2-\alpha_1}\Psi_m(t), \tag{7.88}$$

$$y(t) = K_m^T P_{m\times m}^2\Psi_m(t). \tag{7.89}$$

Since $\Psi_m(t) = \Phi_{m\times m}B_m(t)$, from Eq. (7.89) we have

$$y(t) = K^T P_{m\times m}^2\Phi_{m\times m}B_m(t). \tag{7.90}$$

Let

$$K^T P_{m\times m}^2\Phi_{m\times m} = [a_1, a_2, \ldots, a_m], \tag{7.91}$$

then

$$[y(t)]^3 = \left[a_1^2, a_2^3, \ldots, a_m^3\right] B_m(t). \tag{7.92}$$

TABLE 7.2 Absolute Errors of $y(t)$ with Different Values of m for Example 7.5

t	$m = 24$	$m = 96$	$m = 192$
0.1	8.195145×10^{-5}	5.253765×10^{-6}	1.627688×10^{-6}
0.2	2.051792×10^{-4}	1.265345×10^{-5}	2.515861×10^{-6}
0.3	2.950889×10^{-4}	1.858048×10^{-5}	3.662622×10^{-6}
0.4	3.054479×10^{-4}	1.892447×10^{-5}	6.042111×10^{-6}
0.5	5.080055×10^{-4}	3.171672×10^{-5}	7.927087×10^{-6}
0.6	4.29632×10^{-4}	2.694677×10^{-5}	8.648876×10^{-6}
0.7	6.384631×10^{-4}	3.970229×10^{-5}	8.537701×10^{-6}
0.8	7.117552×10^{-4}	4.459653×10^{-5}	2.75025×10^{-6}
0.9	6.027054×10^{-4}	3.748356×10^{-5}	1.230841×10^{-5}

Similarly, the input signal $f(t)$ may be expanded by the Chebyshev functions as follows:

$$f(t) = f^T \Psi_m(t), \tag{7.93}$$

where f^T is a known constant vector. Substituting Eqs. (7.86)–(7.88) and (7.92) into Eq. (7.83), together with $\Psi_m(t) = \Phi_{m\times m} B_m(t)$ we have

$$\begin{aligned} &K^T \Phi_{m\times m} B_m(t) + K^T P^{2-\alpha_2}_{m\times m} \Phi_{m\times m} B_m(t) B_m(t) \\ &\quad + K^T P^{2-\alpha_1} m \times m \Phi_{m\times m} B_m(t) B_m(t) \left[a_1^3, a_2^3, \ldots, a_m^3\right] B_m(t) \\ &\quad - f^T \Phi_{m\times m} B_m(t) = 0. \end{aligned} \tag{7.94}$$

This is a nonlinear system of algebraic equations. For $a = 1, b = 1, c = 1, e = 1, \alpha_1 = 0.333, \alpha_2 = 1.234$, the Absolute errors of $y(t)$ at given points for different values of m are shown in Table 7.2.

7.3 LEGENDRE WAVELET

Legendre wavelets $\psi_{nm}(t) = \psi(k, \hat{n}, m, t)$ have four arguments; $\hat{n} = 2n - 1, n = 1, 2, 3, ..., 2^{k-1}$, k can assume any positive integer, m is the order for Legendre polynomials and t is the normalized time. They are defined on the interval $[0, 1)$ as follows [55, 94, 108]:

$$\psi_{n,m}(t) = \begin{cases} \sqrt{m + \frac{1}{2}} 2^{k/2} L_m(2^k t - \hat{n}), & \frac{\hat{n}-1}{2^k} \le t < \frac{\hat{n}+1}{2^k}, \\ 0, & \text{otherwise}, \end{cases} \tag{7.95}$$

where $m = 0, 1, 2, \ldots, M - 1, n = 1, 2, \ldots, 2^{k-1}$. The coefficient $\sqrt{m + \frac{1}{2}}$ is for orthogonality. Here, $L_m(t)$ are the well-known Legendre polynomials of order m , which are orthogonal with respect to the weight function $w(t) = 1$ on the interval $[-1, 1]$, and can be determined with the aid of the following recurrence formula:

$$L_0(t) = 1, \qquad L_1(t) = t,$$
$$L_{m+1} = \left(\frac{2m+1}{m+1}\right) tL_m(t) - \left(\frac{m}{m+1}\right) L_{m-1}(t), \quad m = 1, 2, 3, \ldots.$$

The set $\psi_{nm}(t)$ is composed of an orthonormal system for $L^2[0,1]$. A function $y(t)$ defined over $[0,1)$ may be expanded by Legendre wavelet series as

$$y(t) = \sum_{n=1}^{+\infty} \sum_{m=0}^{+\infty} c_{nm}\psi_{nm}(t), \tag{7.96}$$

where

$$c_{nm} = \langle y(t), \psi_{nm}(t)\rangle . \tag{7.97}$$

If the infinite series in Eq. (7.96) is truncated, then Eq. (7.96) can be written as

$$y(t) \simeq \sum_{n=1}^{2^{k-1}} \sum_{m=0}^{M-1} c_{nm}\psi_{nm}(t) = C^T \Psi(t), \tag{7.98}$$

where C and $\Psi(t)$ are $2^{k-1}M \times 1$ matrices given by

$$C(t) = \left[c_{10}, c_{11}, \ldots, c_{1M-1}, c_{20}, \ldots c_{2M-1}, \ldots c_{2^{k-1}0}, \ldots c_{2^{k-1}M-1}\right]^T, \tag{7.99}$$
$$\Psi(t) = \left[\psi_{10}(t), \psi_{11}(t), \ldots, \psi_{1M-1}(t), \ldots, \psi_{2^{k-1}0}(t), \ldots, \psi_{2^{k-1}M-1}(t)\right]^T. \tag{7.100}$$

Taking the collocation points as follows:

$$t_i = \cos\left(\frac{(2i+1)\pi}{2^k M}\right), \quad i = 1, 2, \ldots, 2^{k-1}M. \tag{7.101}$$

We define the Legendre wavelet matrix $\Phi_{m\times m}$ as:

$$\Phi_{m\times m} = [\Psi(t_1), \Psi(t_2), \ldots, \Psi(t_m)], \tag{7.102}$$

where $m = 2^{k-1}M$.

7.3.1 Operational Matrix of Fractional Integration

The integration of the vector $\Psi(t)$ defined in Eq. (7.100) can be obtained as Ref. [55]

$$\int_0^t \Psi(s)ds \simeq P\Psi(t), \tag{7.103}$$

where P is a $2^{k-1}M \times 2^{k-1}M$ matrix given by Ref. [94] as

$$P = \frac{1}{2^k} \begin{bmatrix} L & F & \ldots & F \\ 0 & L & \ldots & F \\ \vdots & 0 & \ddots & \vdots \\ 0 & 0 & \ldots & L \end{bmatrix}, \tag{7.104}$$

where

$$L=\begin{bmatrix} 1 & \frac{\sqrt{3}}{3} & 0 & \cdots & 0 & 0 \\ -\frac{\sqrt{3}}{3} & 0 & \frac{\sqrt{3}}{3\sqrt{5}} & 0 & \cdots & 0 \\ 0 & -\frac{\sqrt{5}}{5\sqrt{3}} & \ddots & \ddots & \ddots & \vdots \\ \vdots & 0 & \ddots & \ddots & \ddots & 0 \\ \vdots & \cdots & \ddots & -\frac{\sqrt{2M-3}}{(2M-3)\sqrt{2M-5}} & \ddots & \frac{\sqrt{2M-3}}{(2M-3)\sqrt{2M-1}} \\ 0 & \cdots & \cdots & 0 & -\frac{\sqrt{2M-1}}{(2M-1)\sqrt{2M-3}} & 0 \end{bmatrix}_{M\times M},$$

and

$$F=\begin{bmatrix} 2 & 0 & \cdots & 0 \\ 0 & 0 & \cdots & 0 \\ \vdots & \vdots & \ddots & \vdots \\ 0 & 0 & \cdots & 0 \end{bmatrix}_{M\times M},$$

The fractional integration of order α of vector $\Psi(t)$ in Eq. (7.100) can be shown as

$$I^{\alpha}\Psi(t) \simeq P^{\alpha}\Psi(t), \tag{7.105}$$

where matrix P^{α} is called Legendre wavelet operational matrix of the fractional integration [55].

The Legendre wavelets can be expanded into m-set of block-pulse functions as

$$\Psi(t) = \Phi_{m\times m}B_m(t), \tag{7.106}$$

where $B_m(t) = [b_0(t)\ b_1(t)\ \ldots\ b_i(t)\ \ldots b_{m-1}(t)]^T$ and $\Phi_{m\times m}$ is defined in Eq. (7.102).

Using Eq. (7.106), we have

$$I^{\alpha}\Psi(t) \approx I^{\alpha}\Phi_{m\times m}B_m(t) = \Phi_{m\times m}I^{\alpha}B_m(t) \approx \Phi_{m\times m}F^{\alpha}B_m(t), \tag{7.107}$$

where F^{α} is defined in Ref. [70]. From Eqs. (7.105) and (7.107), we get

$$P^{\alpha}\Psi(t) = \Phi_{m\times m}F^{\alpha}B_m(t). \tag{7.108}$$

Then, the Chebyshev wavelet operational matrix of the fractional integration $P^{\alpha}_{m\times m}$ is given by

$$P^{\alpha}_{m\times m} = \Phi_{m\times m}F^{\alpha}\Phi^{-1}_{m\times m}. \tag{7.109}$$

7.3.2 Operational Matrix of Product

The following property of the product of two Legendre wavelet vector functions will also be used:

$$\Psi(t)^T\Psi(t)C \simeq \widehat{C}\Psi(t) \tag{7.110}$$

where C is a vector given in Eq. (7.99) and $\widehat{C}$ is a $2^{k-1}M \times 2^{k-1}M$ matrix, which is called the product operation of Legendre wavelet vector functions (see [94]). For $M = 3, k = 1, \widehat{C}$ matrix obtained:

$$\begin{bmatrix} c_{10} & c_{11} & c_{12} \\ c_{11} & c_{10} + \frac{2c_{12}}{\sqrt{5}} & \frac{2c_{11}}{\sqrt{5}} \\ c_{12} & \frac{2c_{11}}{\sqrt{5}} & c_{10} + \frac{2\sqrt{5}c_{12}}{7} \end{bmatrix}.$$

7.3.3 Applications

Example 7.6 *Let us consider the Bagley-Torvik equation [55]*

$$aD^2y(t) + bD^{1.5}y(t) + cy(t) = f(t), \quad t > 0, \tag{7.111}$$

where $f(t) = 1 + t, a = 1, b = 1$ *and* $c = 1$.

Subject to

$$y(0) = y'(0) = 1. \tag{7.112}$$

It is easily verified that the exact solution of this problem is $y(t) = 1 + t$.

We applied the Legendre wavelet approach to solve Eq. (7.111) with $M = 3$ and $k = 1$.

$$D^2y(t) = C^T\Psi(t), \tag{7.113}$$

together with the initial states, then we have

$$D^{1.5}y(t) = (I^{0.5}D^2y)(t) = C^TP^{0.5}\Psi(t), \tag{7.114}$$

$$Dy(t) = C^TP^1\Psi(t) + y'(0), \tag{7.115}$$

$$y(t) = C^TP^2\Psi(t) + y'(0)t + y(0) = C^TP^2\Psi(t) + G^T\Psi(t). \tag{7.116}$$

Similarly, the input signal $f(t)$ may be expanded by the Legendre functions as follows:

$$f(t) = f^T\Psi(t), \tag{7.117}$$

where f^T is a known constant vector. Substituting Eqs. (7.113), (7.114), (7.116) and (7.117) into Eq. (7.111), we get

$$aC^T\Psi(t) + bC^TP^{0.5}\Psi(t) + c(C^TP^2\psi(t) + G^T\Psi(t) = f^T\Psi(t). \tag{7.118}$$

Substituting values of a, b, c into Eq. (7.118) and solving the system of algebraic equations, we can obtain the coefficients C^T. We get $y(t) = 1+t$ which is the exact solution.

Example 7.7 *Consider the following fractional Riccati equation [55]:*

$$D^{\alpha}y(t) = -y(t)^2 + 1, \qquad 0 < \alpha \leq 1, \tag{7.119}$$

subject to initial condition

$$y(0) = 0.$$

The exact solution, when $\alpha = 1$, *is*

$$y(t) = \frac{e^{2t} - 1}{e^{2t} + 1}.$$

We applied the Legendre wavelets approach to solve Eq. (7.119) with $M = 25$ $k = 1$ and various values of α. Figure 7.5 show the approximate solutions obtained for different values of α using the Legendre wavelets scheme. From the graphical results in these figures, it is clear that the approximate solutions converge to the exact solution. Figure 7.6 shows the comparison between the Legendre wavelet solution and the exact solution when $\alpha = 1$.

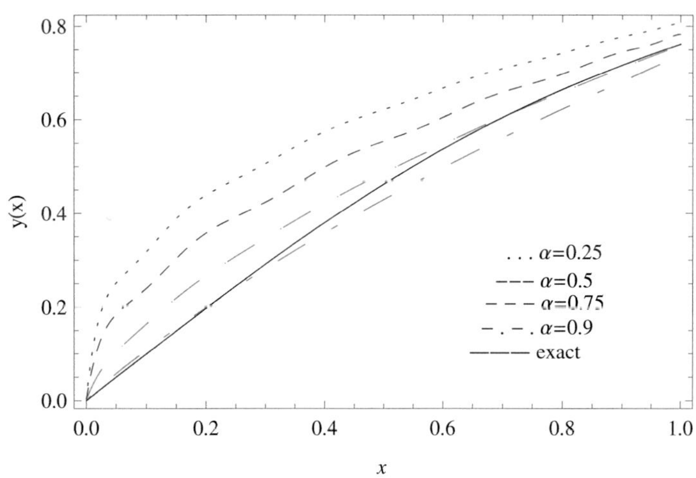

Figure 7.5 Numerical solution using Legendre wavelets when $\alpha = 0.25, 0.5, 0.75, 0.9$ for Example 7.7.

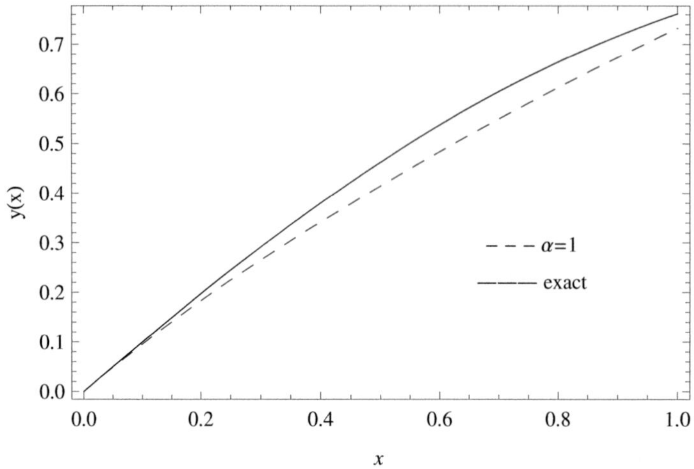

Figure 7.6 Comparison between the Legendre wavelet solution and the exact solution when $\alpha = 1$ for Example 7.7.

Bibliography

[1] Karim Abbaoui and Yves Cherruault. New ideas for proving convergence of decomposition methods. *Computers & Mathematics with Applications*, 29(7):103–108, 1995.

[2] George Adomian. *Solving Frontier Problems of Physics: The Decomposition Method.* Kluwer Academic Publishers, Boston, MA, 1994.

[3] Om Prakash Agrawal. Formulation of Euler-Lagrange equations for fractional variational problems. *Journal of Mathematical Analysis and Applications*, 272(1):368–379, 2002.

[4] Mohammad Nili Ahmadabadi, Farid Mohammad Maalek Ghaini, and M. Arab. Application of He's variational iteration method for Lienard equations. *World Applied Sciences Journal*, 7(9):1077–1079, 2009.

[5] Larry C. Andrews and Bhimsen K. Shivamoggi. *Integral Transforms for Engineers and Applied Mathematicians.* Macmillan, New York, 1988.

[6] Serkan Araci. Novel identities involving Genocchi numbers and polynomials arising from applications of umbral calculus. *Applied Mathematics and Computation*, 233:599–607, 2014.

[7] Aytac Arikoglu and Ibrahim Ozkol. Solution of fractional differential equations by using differential transform method. *Chaos, Solitons & Fractals*, 34(5):1473–1481, 2007.

[8] Ahmed Gamal Atta. Spectral collocation approach with shifted Chebyshev third-kind series approximation for nonlinear generalized fractional Riccati equation. *International Journal of Applied and Computational Mathematics*, 10(2):59, 2024.

[9] Azizollah Babakhani and Varsha Daftardar-Gejji. On calculus of local fractional derivatives. *Journal of Mathematical Analysis and Applications*, 270(1):66–79, 2002.

[10] Ali H. Bhrawy and Abdulaziz Alofi. The operational matrix of fractional integration for shifted Chebyshev polynomials. *Applied Mathematics Letters*, 26(1):25–31, 2013.

[11] Ali H. Bhrawy, Taha M. Taha, and José A. Tenreiro Machado. A review of operational matrices and spectral techniques for fractional calculus. *Nonlinear Dynamics*, 81:1023–1052, 2015.

[12] Ali H. Bhrawy, Mohamed Mahmoud Tharwat, and Mohammad A. Alghamdi. A new operational matrix of fractional integration for shifted Jacobi polynomials. *Bulletin of the Malaysian Mathematical Sciences Society*, 37(4):983–995, 2014.

[13] Jafar Biazar and H. Ghazvini. He's variational iteration method for solving hyperbolic differential equations. *International Journal of Nonlinear Sciences and Numerical Simulation*, 8(3):311–314, 2007.

[14] Aleksei Chechkin, Rudolf Gorenflo, Igor M. Sokolov, and Vsevolod Yu Gonchar. Distributed order time fractional diffusion equation. *Fractional Calculus and Applied Analysis*, 6(3):259–280, 2003.

[15] Theodore S. Chihara. *An Introduction to Orthogonal Polynomials*. Dover Books on Mathematics. Dover Publications, New York, 2011.

[16] H. -W. Choi and J.-G. Shin. Symbolic implementation of the algorithm for calculating Adomian polynomials. *Applied Mathematics and Computation*, 146(1):257–271, 2003.

[17] Varsha Daftardar-Gejji and Azizollah Babakhani. Analysis of a system of fractional differential equations. *Journal of Mathematical Analysis and Applications*, 293(2):511–522, 2004.

[18] Varsha Daftardar-Gejji and Hossein Jafari. Adomian decomposition: A tool for solving a system of fractional differential equations. *Journal of Mathematical Analysis and Applications*, 301(2):508–518, 2005.

[19] Varsha Daftardar-Gejji and Hossein Jafari. An iterative method for solving nonlinear functional equations. *Journal of Mathematical Analysis and Applications*, 316(2):753–763, 2006.

[20] Varsha Daftardar-Gejji and Hossein Jafari. Analysis of a system of nonautonomous fractional differential equations involving Caputo derivatives. *Journal of Mathematical Analysis and Applications*, 328(2):1026–1033, 2007.

[21] Kai Diethelm, Neville J. Ford, and Alan D. Freed. A predictor-corrector approach for the numerical solution of fractional differential equations. *Nonlinear Dynamics*, 29:3–22, 2002.

[22] Kai Diethelm and Alan D. Freed. The fracpece subroutine for the numerical solution of differential equations of fractional order. *Forschung und wissenschaftliches Rechnen*, 1999:57–71, 1998.

[23] Kai Diethelm and Alan D. Freed. On the solution of nonlinear fractional-order differential equations used in the modeling of viscoplasticity. In Frerich Keil, Wolfgang Mackens, Heinrich Voß, and Joachim Werther, editors, *Scientific Computing in Chemical Engineering II*, pp. 217–224. Springer, Berlin Heidelberg, 1999.

[24] Kai Diethelm. Generalized compound quadrature formulae for finite-part integrals. *IMA Journal of Numerical Analysis*, 17(3):479–493, 1997.

[25] Peng Ding, Yubin Yan, Zongqi Liang, and Yuyuan Yan. Finite difference method for time-fractional Klein–Gordon equation on an unbounded domain using artificial boundary conditions. *Mathematics and Computers in Simulation*, 205:902–925, 2023.

[26] John T. Edwards, Neville J. Ford, and A. Charles Simpson. The numerical solution of linear multi-term fractional differential equations: Systems of equations. *Journal of Computational and Applied Mathematics*, 148(2):401–418, 2002.

[27] Nader Engheta. On the role of fractional calculus in electromagnetic theory. *IEEE Antennas and Propagation Magazine*, 39(4):35–46, 1997.

[28] Nader Engheta. Fractional duality in electromagnetic theory. In *Proceedings of the URSI International Symposium on Electromagnetic Theory*, Proc. URSI International Symposium on Electromagnetic Theory Salonique, 575-577, 1998.

[29] Vedat Suat Ertürk and Shaher Momani. Solving systems of fractional differential equations using differential transform method. *Journal of Computational and Applied Mathematics*, 215(1):142–151, 2008.

[30] Zhi-Wei Fang, Jia-Li Zhang, and Hai-Wei Sun. A fast finite volume method for spatial fractional diffusion equations on nonuniform meshes. *Computers & Mathematics with Applications*, 108:175–184, 2022.

[31] Roghayeh Moallem, Hossein Jafari, Maluti Kgarose, and Seyed Abolghasem Mohammadi. Numerical solutions of time-fractional Klein-Gordon equations by Clique polynomials. *Alexandria Engineering Journal*, 60(5):4563–4571, 2021.

[32] Yusif S. Gasimov and Juan Eduardo Napoles-Vald. Some refinements of hermite-hadamard inequality using k-fractional Caputo derivatives. *Fractional Differential Calculus*, 12(2):209–21, 2022.

[33] Asghar Ghorbani and Jafar Saberi-Nadjafi. He's homotopy perturbation method for calculating Adomian polynomials. *International Journal of Nonlinear Sciences and Numerical Simulation*, 8(2):229–232, 2007.

[34] Alireza Khalili Golmankhaneh, Kerri Welch, Cemil Tunç, and Yusif S. Gasimov. Classical mechanics on fractal curves. *The European Physical Journal Special Topics*, 232(7):991–999, 2023.

[35] Jaideva C. Goswami and Andrew K. Chan. *Fundamentals of Wavelets: Theory, Algorithms, and Applications*. John Wiley & Sons, Hoboken, NJ, 2011.

[36] F. Alberto Grünbaum. Matrix valued Jacobi polynomials. *Bulletin des Sciences Mathématiques*, 127(3):207–214, 2003.

[37] Emanuel Guariglia and Rodrigo Capobianco Guido. Chebyshev wavelet analysis. *Journal of Function Spaces*, 2022(1):5542054, 2022.

[38] Ahmad Haghbin, Hossein Jafari, Pranay Goswami, and Morganathan Vernon Ariyan. Solving time-fractional chemical engineering equations by generalized differential transform method. *Thermal Science*, 24(Suppl. 1):157–164, 2020.

[39] Ji-Huan He. Non-perturbative methods for Strongly nonlinear problems. Wissenschaft & Technik. dissertation.de - Verlag im Internet GmbH, 2006.

[40] Ji-Huan He. Approximate analytical solution for seepage flow with fractional derivatives in porous media. *Computer Methods in Applied Mechanics and Engineering*, 167(1):57–68, 1998.

[41] Ji-Huan He. Homotopy perturbation technique. *Computer Methods in Applied Mechanics and Engineering*, 178(3):257–262, 1999.

[42] Ji-Huan He. Variational iteration method - A kind of non-linear analytical technique: Some examples. *International Journal of Non-Linear Mechanics*, 34(4):699–708, 1999.

[43] Ji-Huan He. Homotopy perturbation method: A new nonlinear analytical technique. *Applied Mathematics and Computation*, 135(1):73–79, 2003.

[44] Ji-Huan He and Xu-Hong Wu. Variational iteration method: New development and applications. *Computers & Mathematics with Applications*, 54(7):881–894, 2007.

[45] Rudolf Hilfer. *Applications of Fractional Calculus in Physics*. World Scientific Publishing Company, Farrer Road, Singapore, 912805, 2000.

[46] Chunhui Hsiao and S. P. Wu. Numerical solution of time-varying functional differential equations via Haar wavelets. *Applied Mathematics and Computation*, 188(1):1049–1058, 2007.

[47] M. Idrees Bhatti and Paul Francis Bracken. Solutions of differential equations in a Bernstein polynomial basis. *Journal of Computational and Applied Mathematics*, 205(1):272–280, 2007.

[48] Mitio Inokuti, Hideki Sekine, and Toshio Mura. General use of the lagrange multiplier in nonlinear mathematical physics. *Proceedings of the IUTAM Symposium on Variational Methods in the Mechanics of Solids*, Northwestern University, Evanston, IL, 1980.

[49] Hossein Jafari. Iterative methods for solving system of fractional differential equations. PhD thesis, Pune University Pune, India, 2006.

[50] Hossein Jafari, Chaudry Masood Khalique, and Marjan Nazari. An algorithm for the numerical solution of nonlinear fractional-order Van der Pol oscillator equation. *Mathematical and Computer Modelling*, 55(5):1782–1786, 2012.

[51] Hossein Jafari and Samaneh Seifi. Homotopy analysis method for solving linear and nonlinear fractional diffusion-wave equation. *Communications in Nonlinear Science and Numerical Simulation*, 14(5):2006–2012, 2009.

[52] Hossein Jafari and Haleh Tajadodi. He's variational iteration method for solving fractional Riccati differential equation. *International Journal of Differential Equations*, 2010(1):764738, 2010.

[53] Hossein Jafari, Haleh Tajadodi, and Seyed Ali Hosseini Matikolai. Homotopy perturbation pade technique for solving fractional Riccati differential equations. *International Journal of Nonlinear Sciences and Numerical Simulation*, 11(Supplement):271–276, 2010.

[54] Hossein Jafari, Haleh Tajadodi, Marjan Nazari, and Chaudry Masood Khalique. Numerical solution of non-linear Riccati differential equations with fractional order. *International Journal of Nonlinear Sciences and Numerical Simulation*, 11(Supplement):179–182, 2010.

[55] Hossein Jafari, Sohrabali Yousefi, Mohammad Arab Firoozjaee, Shaher Momani, and Chaudry Masood Khalique. Application of Legendre wavelets for solving fractional differential equations. *Computers & Mathematics with Applications*, 62(3):1038–1045, 2011. Special Issue on Advances in Fractional Differential Equations II.

[56] Hossein Jafari, Maryam Alipour, and Hale Tajadodi. Two-dimensional differential transform method for solving nonlinear partial differential equations. *International Journal of Research and Reviews in Applied Sciences*, 2(1):47–52, 2010.

[57] Hossein Jafari and Varsha Daftardar-Gejji. Solving a system of nonlinear fractional differential equations using Adomian decomposition. *Journal of Computational and Applied Mathematics*, 196(2):644–651, 2006.

[58] Hossein Jafari and Varsha Daftardar-Gejji. Solving linear and nonlinear fractional diffusion and wave equations by Adomian decomposition. *Applied Mathematics and Computation*, 180(2):488–497, 2006.

[59] Hossein Jafari, Roghayeh Moallem Ganji, Davood Domiri Ganji, Zakia Hammouch, and Yusif S. Gasimov. A novel numerical method for solving Fuzzy variable-order differential equations with Mittag-Leffler kernels. *Fractals*, 31(04):2340063, 2023.

[60] Hossein Jafari, Roghayeh Moallem Ganji, Sonali Mandar Narsale, Maluti Kgarose, and Van Thinh Nguyen. Application of Hosoya polynomial to solve a class of time-fractional diffusion equations. *Fractals*, 31(04):2340059, 2023.

[61] Hossein Jafari, Saber Ghasempoor, and Chaudry Masood Khalique. A comparison between Adomian's polynomials and He's polynomials for nonlinear functional equations. *Mathematical Problems in Engineering*, 2013(1):943232, 2013.

[62] Hossein Jafari, Babak Fazli Malidareh, and Vahid Reza Hosseini. Collocation discrete least squares meshless method for solving nonlinear multi-term time fractional differential equations. *Engineering Analysis with Boundary Elements*, 158:107–120, 2024.

[63] Hossein Jafari and Shaher Momani. Solving fractional diffusion and wave equations by modified homotopy perturbation method. *Physics Letters A*, 370(5):388–396, 2007.

[64] Hossein Jafari, Hale Tajadodi, and Dumitru Baleanu. A modified variational iteration method for solving fractional Riccati differential equation by Adomian polynomials. *Fractional Calculus and Applied Analysis*, 16(1):109–122, 2013.

[65] Hossein Jafari and Haleh Tajadodi. Fractional order optimal control problems via the operational matrices of Bernstein polynomials. *UPB Scientific Bulletin*, 76(3):115–128, 2014.

[66] Hossein Jafari and Haleh Tajadodi. Electro-spunorganic nanofibers elaboration process investigations using BPS operational matrices. *Iranian Journal of Mathematical Chemistry*, 7(1):19–27, 2016.

[67] Doğan Kaya and Salah M. El-Sayed. A numerical solution of the Klein-Gordon equation and convergence of the decomposition method. *Applied Mathematics and Computation*, 156(2):341–353, 2004.

[68] Saeed Kazem. An integral operational matrix based on Jacobi polynomials for solving fractional-order differential equations. *Applied Mathematical Modelling*, 37(3):1126–1136, 2013.

[69] Hasib Khan, Hossein Jafari, Rahmat Ali Khan, Haleh Tajadodi, and Sarah Jane Johnston. Numerical solutions of the nonlinear fractional-order Brusselator system by Bernstein polynomials. *The Scientific World Journal*, 2014(1):257484, 2014.

[70] Adem Kilicman and Zeyad Abdel Aziz Al Zhour. Kronecker operational matrices for fractional calculus and some applications. *Applied Mathematics and Computation*, 187(1):250–265, 2007. *Proceedings of the International Symposium on Analytic Function Theory, Fractional Calculus and Their Applications in Honour of Professor H.M. Srivastava on his Sixty-Fifth Birth Anniversary.*

[71] Kiran M. Kolwankar and Anil D. Gangal. Fractional differentiability of nowhere differentiable functions and dimensions. *Chaos: An Interdisciplinary Journal of Nonlinear Science*, 6(4):505–513, 12 1996.

[72] John Denholm Lambert. *Numerical Methods for Ordinary Differential Systems: The Initial Value Problem.* Wiley, Hoboken, NJ, 1991.

[73] Yuanlu Li. Solving a nonlinear fractional differential equation using Chebyshev wavelets. *Communications in Nonlinear Science and Numerical Simulation*, 15(9):2284–2292, 2010.

[74] Yuanlu Li and Weiwei Zhao. Haar wavelet operational matrix of fractional order integration and its applications in solving the fractional order differential equations. *Applied Mathematics and Computation*, 216(8):2276–2285, 2010.

[75] Shijun. *Beyond Perturbation: Introduction to the Homotopy Analysis Method.* Chapman & Hall, CRC Press, New York, 2003.

[76] Ch. Lubich. Discretized fractional calculus. *SIAM Journal on Mathematical Analysis*, 17(3):704–719, 1986.

[77] Min Ma, Dumitru Baleanu, Yusif S. Gasimov, and Xiao-Jun Yang. New results for multidimensional diffusion equations in fractal dimensional space. *Romanian Journal of Physics*, 61:784–794, 2016.

[78] Benoit B. Mandelbrot. *The Fractal Geometry of Nature.* W.H. Freeman, San Francisco, CA, 1982.

[79] John C. Mason and David C. Handscomb. *Chebyshev Polynomials.* Chapman & Hall/CRC, New York, 2002.

[80] Yves Meyer and Sylvie Roques. *Progress in Wavelet Analysis and Applications: Proceedings of the International Conference "Wavelets and Applications," Toulouse, France - June 1992.* Editions Frontieres, Atlantica Séguier Frontières.

[81] Kenneth S. Miller and Bertram Ross. *An Introduction to the Fractional Calculus and Fractional Differential Equations.* John Wiley & Sons, Inc., New York, 1993.

[82] Shaher Momani and Zaid Odibat. Numerical comparison of methods for solving linear differential equations of fractional order. *Chaos, Solitons & Fractals*, 31(5):1248–1255, 2007.

[83] Shaher Momani, Zaid Odibat, and Ahmed Alawneh. Variational iteration method for solving the space- and time-fractional KdV equation. *Numerical Methods for Partial Differential Equations*, 24(1):262–271, 2008.

[84] Sonali Mandar Narsale, Ram Kishun Lodhi, and Hossein Jafari. A numerical study for non-linear multi-term fractional order differential equations. *Advanced Mathematical Models & Applications*, 9(2):193–204, 2024.

[85] Katsuyuki Nishimoto. *An Essence of Nishimoto's Fractional Calculus (Calculus in the 21st Century): Integrations and Differentiations of Arbitrary Order.* Descartes Press Company, Koriyama, 1991.

[86] Zaid Odibat and Shaher Momani. Numerical methods for nonlinear partial differential equations of fractional order. *Applied Mathematical Modelling*, 32(1):28–39, 2008.

[87] Zaid Odibat and Shaher Momani. The variational iteration method: An efficient scheme for handling fractional partial differential equations in fluid mechanics. *Computers & Mathematics with Applications*, 58(11):2199–2208, 2009. New Analytical Methods and Applications.

[88] Zaid Odibat and Shaher Momani. Application of variational iteration method to nonlinear differential equations of fractional order. *International Journal of Nonlinear Sciences and Numerical Simulation*, 7(1):27–34, 2006.

[89] Abhay Parvate and Anil D. Gangal. Fractal differential equations and fractal-time dynamical systems. *Pramana*, 64:389–409, 2005.

[90] Peter C. B. Phillips. *Fractional Matrix Calculus and the Distribution of Multivariate Tests*, pp. 219–234. Springer, Netherlands, Dordrecht, 1987.

[91] Peter C. B. Phillips. The exact distribution of the wald statistic. *Econometrica: Journal of the Econometric Society*, 54(4):881–895, 1986.

[92] Igor Podlubny. *Fractional Differential Equations: An Introduction to Fractional Derivatives, Fractional Differential Equations, to Methods of Their Solution and some of Their Applications.* Mathematics in Science and Engineering. Elsevier Science, San Diego, California, 1998.

[93] Pankaj Ramani, Arif M. Khan, Daya Lal Suthar, and Dinesh Kumar. Approximate analytical solution for non-linear Fitzhugh–Nagumo equation of time fractional order through fractional reduced differential transform method. *International Journal of Applied and Computational Mathematics*, 8(2):61, 2022.

[94] Mohsen Razzaghi and Sohrabali Yousefi. The legendre wavelets operational matrix of integration. *International Journal of Systems Science*, 32(4):495–502, 2001.

[95] Fred Riewe. Nonconservative Lagrangian and Hamiltonian mechanics. *Physical Review E*, 53:1890–1899, 1996.

[96] Fred Riewe. Mechanics with fractional derivatives. *Physical Review E*, 55:3581–3592, 1997.

[97] Fariba Rigi and Haleh Tajadodi. Numerical approach of fractional abel differential equation by Genocchi polynomials. *International Journal of Applied and Computational Mathematics*, 5(5):134, 2019.

[98] Abbas Saadatmandi. Bernstein operational matrix of fractional derivatives and its applications. *Applied Mathematical Modelling*, 38(4):1365–1372, 2014.

[99] Nikita Saha and Randhir Singh. An efficient new numerical algorithm for solving Emden–Fowler pantograph differential equation using Laguerre polynomials. *Journal of Computational Science*, 72:102108, 2023.

[100] Saha Salati, Mashallah Matinfar, and Hossein Jafari. A numerical approach for solving Bagely-Torvik and fractional oscillation equations. *Advanced Mathematical Models & Applications*, 8(2):241–52, 2023.

[101] Saha Salati, Hossein Jafari, and Mashallah Matinfar. A numerical method for solving a class of variable-order differential equations using Hosoya polynomial of simple paths. *Journal of Applied Mathematics and Computational Mechanics*, 23(1):97–108, 2024.

[102] Mahmood Sheybak and Haleh Tajadodi. Numerical solutions of fractional chemical kinetics system. *Nonlinear Dynamics and Systems Theory*, 19(1):200–208, 2019.

[103] Harendra Singh and Ramta Ram Pathak. Jacobi spectral method for the fractional reaction–diffusion equation arising in ecology. *Mathematical Methods in the Applied Sciences*, 47(6):5031–5045, 2024.

[104] Vineet K. Srivastava, Mukesh K. Awasthi, and Mohammad Tamsir. RDTM solution of Caputo time fractional-order hyperbolic Telegraph equation. *AIP Advances*, 3(3):032142, 2013.

[105] Haleh Tajadodi, Hasib Khan, Jehad Alzabut, and José Francisco Gómez-Aguilar. An optimization method for solving fractional oscillation equation. *Results in Physics*, 57:107403, 2024.

[106] Mehdi Tatari and Mehdi Dehghan. Improvement of he's variational iteration method for solving systems of differential equations. *Computers & Mathematics with Applications*, 58(11):2160–2166, 2009. New Analytical Methods and Applications.

[107] Majid Tavassoli Kajani, Abdollah Hadi Vencheh, and Mehdi Ghasemi. The Chebyshev wavelets operational matrix of integration and product operation matrix. *International Journal of Computer Mathematics*, 86(7):1118–1125, 2009.

[108] Bharti Thakur and Sandipan Gupta. An iterative algorithm for numerical solution of nonlinear fractional differential equation using Legendre wavelet method. *IAENG International Journal of Applied Mathematics*, 54(3):380–389, 2024.

[109] Yu Wang and Min Cai. A fast finite difference scheme for the time-space fractional diffusion equation. *IFAC-PapersOnLine*, 58(12):174–178, 2024. *12th IFAC Conference on Fractional Differentiation and its Applications ICFDA 2024.*

[110] Abdul-Majid Wazwaz. A reliable modification of Adomian decomposition method. *Applied Mathematics and Computation*, 102(1):77–86, 1999.

[111] Abdul-Majid Wazwaz. The variational iteration method for solving linear and nonlinear systems of PDEs. *Computers & Mathematics with Applications*, 54(7):895–902, 2007. Variational Iteration Method for Nonlinear Problems.

[112] Abdul-Majid Wazwaz. *A First Course in Integral Equations.* World Scientific, 5 Toh Tuch Link, Singapore, 596224, 2nd edition, 2015.

[113] Bruce West, Mauro Bologna, and Paolo Grigolini. *Physics of Fractal Operators*. Institute for Nonlinear Science. Springer, Berlin, Germany, 2003.

[114] Rahat Zarin, Hammad Khaliq, Amir Khan, Iftikhar Ahmed, and Usa Wannasingha Humphries. A numerical study based on Haar wavelet collocation methods of fractional-order antidotal computer virus model. *Symmetry*, 15(3), 2023.

[115] Tie Zhang and Qingxin Guo. The finite difference/finite volume method for solving the fractional diffusion equation. *Journal of Computational Physics*, 375:120–134, 2018.

[116] J. K. Zhou. *Differential Transformation and Its Applications for Electrical Circuits*. Huazhong University Press, Wuhan, China, 1986.

[117] Ping Zhou, Hossein Jafari, Roghayeh Moallem Ganji, and Sonali M. Narsale. Numerical study for a class of time fractional diffusion equations using operational matrices based on Hosoya polynomial. *Electronic Research Archive*, 31(8):4530–4548, 2023.

[118] Canan Ünlü, Hossein Jafari, and Dumitru Baleanu. Revised variational iteration method for solving systems of nonlinear fractional-order differential equations. *Abstract and Applied Analysis*, 2013(1):461837, 2013.

[119] Şuayip Yüzbaşı. A collocation method based on Bernstein polynomials to solve nonlinear Fredholm-Volterra integro-differential equations. *Applied Mathematics and Computation*, 273:142–154, 2016.

Index